Andreas Pawlak
Advanced Modeling of Silicon-Germanium
Heterojunction Bipolar Transistors

TUD*press*

Andreas Pawlak

Advanced Modeling of Silicon-Germanium Heterojunction Bipolar Transistors

TUDpress

2015

Die vorliegende Arbeit wurde am 01. Dezember 2014 an der Fakultät
Elektrotechnik und Informationstechnik der Technischen Universität Dresden
als Dissertation eingereicht und am 05. Februar 2015 verteidigt.

Vorsitzender:
Prof. Dr.-Ing. habil. Jörges

Gutachter:
Prof. Dr.-Ing. habil. Schröter
Prof. Dr.-Ing. Möller

Bibliografische Information der Deutschen Nationalbibliothek
Die Deutsche Nationalbibliothek verzeichnet diese Publikation in der
Deutschen Nationalbibliografie; detaillierte bibliografische Daten sind
im Internet über http://dnb.d-nb.de abrufbar.

Bibliographic information published by the Deutsche Nationalbibliothek
The Deutsche Nationalbibliothek lists this publication in the Deutsche
Nationalbibliografie; detailed bibliographic data are available in the
Internet at http://dnb.d-nb.de.

ISBN 978-3-95908-028-6

© 2015 TUDpress
Verlag der Wissenschaften GmbH
Bergstr. 70 | D-01069 Dresden
Tel.: 0351/47 96 97 20 | Fax: 0351/47 96 08 19
http://www.tudpress.de

Technische Universität Dresden

Advanced Modeling of Silicon-Germanium Heterojunction Bipolar Transistors

Andreas Pawlak

von der Fakultät Elektrotechnik und Informationstechnik der Technischen Universität Dresden

zur Erlangung des akademischen Grades

Doktoringenieur

(Dr.-Ing.)

genehmigte Dissertation

Vorsitzender: Prof. Dr.-Ing. habil. Jörges
Gutachter: Prof. Dr.-Ing. habil. Schröter   Tag der Einreichung: 01.12.2014
Prof. Dr.-Ing. Möller   Tag der Verteidigung: 05.02.2015

*Für Irina
und Daria*

Danksagung

Der wichtigste Dank für diese Arbeit gebührt Prof. Michael Schröter, der mit seinen Visionen und als Doktorvater die wichtigsten Grundsteine für diese Arbeit gelegt hat. Ohne sein wissenschaftliches Verständnis, seine Betreuung und die Hilfsmittel, welche er zur Verfügung gestellt hat, und letztendlich natürlich sein Vertrauen in mich, welches zur Anstellung an seinem Lehrstuhl geführt hat, wären die Ergebnisse dieser Arbeit niemals zustande gekommen.

Ein herzlicher Dank geht auch an Prof. Michael Möller für die Beurteilung dieser Dissertation und die Zeit, die er dafür investiert hat.

Ein besonderer Dank richtet sich an den gesamten Lehrstuhl für Elektronische Bauelemente und Integrierte Schaltungen an der Technischen Universität Dresden. Allen voran Julia Krause als beste "Raumteilerin" überhaupt, die in stressigen Zeiten immer die richtigen motivierenden Worte gefunden hat, und Steffen Lehmann, der als mein Betreuer zu Studentenzeiten die Grundsteine gelegt hat und sich auch später immer Zeit für jedwede fachliche und nicht-fachliche Diskussion genommen hat. Weiterhin danke ich Gerald Wedel für seine fantastische Unterstützung bei Bauelementesimulationen und der Erweiterung meines "kompaktmodellischen" Weltbildes, sowie Kai-Erik Moebus, Martin Claus, Yves Zimmermann, Holger Wittkopf, Tommy Rosenbaum und Tobias Nardmann für die immer sehr angeregten Diskussionen auch mal über Bipolartransistoren hinaus. Weiterhin zu nennen sind Anindya Mukherjee, der dafür gesorgt hat, dass ich die Bedürfnisse eines Schaltungsentwicklers besser einschätzen kann, sowie Paulius Sakalas für die Bereitstellung zahlreicher Messergebnisse und seine Unterstützung im Labor. Letztendlich gebührt natürlich auch unserer Sekretärin Ria Lykowski ein riesiger Dank dafür, dass sie uns die alltägliche Arbeit an einer Universität so sehr erleichtert sowie allen weiteren Mitarbeitern des Lehrstuhls für ihre Unterstützung. Ein besonderer Dank geht noch einmal an Kai, Julia, Steffen, Tobias, Yves und Gerald für das Korrigieren der Arbeit.

Weiterhin danke ich Dr. Bernd Heinemann, Dr. Gerhard Fischer, Dr. Alexander Fox, Dr. Holger Rücker, Dr. René Scholz und Christian Wipf (alle Mitarbeiter am IHP) für die sehr gute Zusammenarbeit mit dem Institut. Vielen Dank an Jörg Berkner von Infineon für die herzliche Aufnahme im Bipolar Arbeitskreis und die mit ihm, Franz Sischka und Zoltan Huszka geführten fachlichen Diskussionen.

I also want to thank the staff of ST Microelectronics, especially Didier Céli who always was a driving force for improvements of the compact model and extraction routines. Additional thanks are directed to the beautiful city of Bordeaux and the team of XMOD and the IMS there.

Einen besonderen Dank möchte ich an meine gesamte Familie richten, ohne die diese Arbeit niemals zustande gekommen wäre. Der Dank richtet sich insbesondere an meine Frau Irina, die mich zu jeder Zeit unterstützt hat und mir die notwendige Kraft für diese Arbeit gegeben hat, sowie an meine Tochter Daria, die stark war und es akzeptiert hat, dass Papa zum Ende der Arbeit nur sehr wenig Zeit mit ihr verbringen konnte. Ein besonderer Dank geht an meine Mutter Beate dafür, dass sie mir nicht nur eine tolle Kindheit, sondern auch das Studium, welches die Grundlage für diese Arbeit war, ermöglicht hat.

Kurzfassung

Schnelle Silizium-Germanium (SiGe) Hetero-Bipolartransistoren (HBTs) sind hervorragend für die Anwendung in Schaltungen geeignet, welche den Frequenzbereich von 0.3 bis ca. 2 THz (sog. Terahertz Gap) abdecken. Da die Kosten für einen Entwurfszyklus jedoch mit steigenden Frequenzen und höherer Komplexität der Systeme deutlich ansteigen, sind genaue physikalisch basierte Kompaktmodelle für den Schaltungsentwurf unverzichtbar. In dieser Dissertation werden daher spezielle physikalische Effekte von schnellen SiGe-Transistoren betrachtet, welche noch nicht oder nur unzureichend in bisherigen Kompaktmodellen berücksichtigt wurden.

Das durch das Einbringen von Germanium in die Basis eines Silizium-Bipolartransistors entstehende nicht-ideale Verhalten des Transferstroms wird untersucht. Die physikalische Hintergründe dieser Effekte werden beschrieben und darauf basierend Kompaktmodellgleichungen entwickelt. Die Grundlage für die Modellierung ist dabei die "Generalized Integral Charge Control Relation" (GICCR), welche den Kern der Transferstromberechnung im verwendeten Kompaktmodell HICUM Level 2 (HICUM/L2) bildet. Die entwickelten Modellgleichungen wurden erfolgreich auf verschiedene SiGe-Technologien angewandt.

Während der Untersuchung der linearen und nicht-linearen Modellierung von Bipolartransistoren mit HICUM/L2 wurden Ungenauigkeiten bei der Beschreibung der Kleinsignalgrößen aufgezeigt. Als Ursache dieser Probleme wurde das Modell für die nicht-quasi-statischen (NQS) Effekte sowie die fehlende Modellierung der Stromabhängigkeit der Kollektorladung identifiziert. Für diese Stromabhängigkeit wurde ein physikalisch basiertes Modell auf Grundlage von messbaren Kleinsignalgrößen entwickelt. Es wurden sehr gute Ergebnisse bei der Anwendung auf numerische Simulationen erzielt. Bei der Anwendung auf Messungen zeigten sich allerdings noch kleinere Schwächen im Modell.

In einem kurzen Abschnitt über die Modellierung der Substratkopplung wird ein neues Modell für die Arbeitspunktabhängigkeit des Substratwiderstands entwickelt sowie die geeignete Aufteilung der Substratkapazität gezeigt.

Darüber hinaus werden sowohl neue Extraktionsmethoden für Kompaktmodellparameter anhand von Standardkennlinien vorgestellt, als auch existierende Methoden hinsichtlich ihrer Anwendbarkeit auf moderne Prozesse untersucht. Die entwickelten Extraktionsmethoden für den Emitter- und thermischen Widerstand sowie für die Beschreibung des Transferstroms für Bauelemente unterschiedlicher Emittergeometrie wurden erfolgreich auf schnelle Bipolartransistoren angewendet. Die allgemeine bilineare Skalierungsgleichung ermöglicht die einfache Beschreibung von nicht ideal skalierenden Strömen. Die Anwendung ist dabei nicht auf statische Ströme begrenzt. Sie ist auch für die Beschreibung dynamischer Ströme möglich.

Den Abschluss der Dissertation bildet die Anwendung der entwickelten Modellgleichungen auf einen der modernsten schnellen Bipolarprozesse. Hierfür wurde die SG13G2 Technologie vom IHP gewählt. Anhand der sehr guten Modellierung des DC- und Kleinsignal- sowie des Großsignalbetriebs werden die Verbesserungen des Kompaktmodells HICUM/L2, welche im Rahmen dieser Arbeit entstanden sind, demonstriert.

Abstract

Advanced Silicon-Germanium (SiGe) hetero-junction bipolar-transistors (HBTs) are perfectly suited for circuit applications targeting the frequency range from approximately 0.3 to 2 THz, the so-called terahertz gap. Since the fabrication costs per design cycle are rapidly increasing with progressing frequency and complexity of the systems, accurate compact models are essential in order to enable robust circuit design. This work focuses on selected important physical effects in advanced SiGe-transistors, which have been either insufficiently modeled or completely missing in conventional compact models.

The physical origin of several non-ideal effects for the transfer current caused by the introduction of Germanium into the base of a silicon bipolar transistor is discussed. New compact model equations were derived and successfully applied to a large set of different technologies. Hereby, the "Generalized Integral Charge Control Relation", which is the key concept for the description of the transfer current of the employed compact model HICUM level 2 (HICUM/L2), was used as a foundation.

Based on an initial evaluation of the linear and non-linear modeling capabilities of HICUM/L2, model related issues of the small-signal parameters are discussed. Both the implemented non-quasi-static (NQS) model and the missing model for the current dependent collector charge are shown to be the origin of those issues. For the latter a physics-based model utilizing small-signal parameters obtained from measurements is derived. Although the model has been successfully verified based on numerical device simulations, it is only partially accurate when compared to experimental results.

A brief chapter about substrate effects in bipolar transistors comprises the derivation of a compact model for the bias-dependent substrate resistance as well as a proper partitioning of the substrate capacitance.

Furthermore, new extraction methods for compact model parameters are introduced. In addition, the application of existing methods to advanced processes is discussed. The derived joint extraction method for the emitter and thermal resistance as well as a scalable model for the transfer current have been successfully applied to experimental data of fast HBTs. Hereby, the general bi-linear scaling approach allows the modeling of non-ideally scaling currents. Moreover, the application is not limited to static currents but can also be applied to dynamic currents.

Finally, the derived model equations were applied to a selected very advanced SiGe HBT process developed by IHP. Highly accurate models for DC- and small-signal as well as for large-signal characteristics are presented. Hence, this work significantly improves the state-of-the-art in SiGe HBT compact modeling.

Table of Contents

Often used symbols and acronyms

β_f	DC current gain
V_{CE}, V_{SC}	Terminal collector-emitter and substrate-collector voltage
B, C, E, S	Base, collector, emitter and substrate (contact)
BE, BC	Base-emitter, base-collector
BiCMOS	Bipolar CMOS
BJT	Bipolar junction transistor
C_{BE}, C_{BC}	Terminal BE- and BC-capacitance
C_{dEi}, C_{dCi}	BE- and BC-diffusion capacitance
C_{jEi}, C_{jCi}	BE- and BC-depletion capacitance
CE, SC	Base-emitter, base-collector
CMOS	Complementary MOS
CNTFET	carbon nanotube field-effect transistor
DTI	Deep trench isolation
E_{jC}	Magnitude of the electric field in the BC-SCR
EC	Equivalent circuit
f_0	Fundamental frequency
f_T, f_{max}	Cutoff frequency and maximum oscillation frequency
FOM	Figure of merit
G_T, PAE	Transducer gain and power added efficiency
h_0, h_{f0}	Weight factor the for zero-bias charge and the low injection mobile charge

h_{BE}, h_{BC}	Weight factor for the BE- and BC-depletion charge
h_{fE}, h_{fC}	Normalized weight factor for the mobile emitter and collector charge
h_{jEi}, h_{jCi}	Normalized weight factor for the BE- and BC-depletion charge
h_{mE}, h_{mBE}, h_{mB}	Weight factor for the mobile charge in the emitter, BE-SCR and the base
h_{pBC}	Weight factor for the collector charge
HBT	Heterojunction bipolar transistor
HICUM	Bipolar transistor compact model, High Current Model
I_{BEs}, I_{BCs}, I_{SCs}	Saturation currents of the base-emitter, base-collector and substrate-collector diode
I_{BE}, I_{BC}, I_{SC}	Currents of the base-emitter, base-collector and substrate-collector diode
I_B, I_C, I_E, I_S	Terminal base, collector, emitter and substrate current
II	Impact ionization
InP	Indium phosphide
N_B, N_C, N_E	Doping concentration in the base, collector and emitter of a transistor
NBR	Neutral base recombination
NQS	Non-quasi static
P_{avs}, P_{out}	Maximum available power and output power
Q_{jEi}, Q_{jCi}	BE- and BC-depletion charge
Q_{p0}, Q_p, $Q_{p,T}$	(Zero-bias) hole charge, weighted hole charge
R_B, R_{Bi}, R_{Bx}	Complete base resistance, internal and external base resistance
R_{Cx}, R_E	External collector and emitter series resistance
r_{KB}, r_{KC}	Length specific base and collector contact resistance
r_{sBi}	Internal base sheet resistance
r_{sBL}	Buried layer sheet resistance

r_{sPM}, r_{sPO}	Non-silicided poly over mono and poly over oxide sheet resistance
r_{sSp}, r_{sSil}	Base sheet resistance under to spacer, silicided base sheet resistance
R_{th}	Thermal resistance
SCR	Space charge region
Si, Ge, SiGe	Silicon, Germanium, Silicon-germanium
STI	Shallow trench isolation
T_0, T_{amb}, T, ΔT	Reference temperature, ambient temperature, actual temperature and increase of the actual temperature over the ambient temperature
TCAD	Technology computer aided design
$V_{B'E'}$, $V_{B'C'}$	Internal base-emitter and base-collector voltage
$V_{B*E'}$	Peripheral base-emitter voltage
V_{BE}, V_{BC}	Terminal base-emitter and base-collector voltage
$V_{C'E'}$, $V_{S'C'}$	Internal collector-emitter and substrate-collector voltage
v_{sn}, v_{sp}	Electron and hole saturation velocity
V_T	Thermal voltage
w_{B0}, w_B	(Zero-voltage) base width
x_E, x_C	End of the neutral base at the emitter and collector side
x_{jE0}, x_{jC0}	Location of the metallurgic base-emitter and base-collector junction at zero volt
x_{jE}, x_{jC}	Location of the metallurgic base-emitter and base-collector junction

1 | Introduction

Integrated silicon-germanium heterojunction bipolar transistors (SiGe HBTs) are one of the key technologies for analog applications due to their high speed and high driving capabilities. They enable the design of high-speed RF front-ends at low DC power for advanced process generations.

State-of-the-art transistors with (f_T/f_{max}) up to $(300,500)$ GHz have been reported ([CPL$^+$09,HBB$^+$10,CCLA$^+$13]). This operating speed is currently increased by ongoing projects as, e.g., the DOTSEVEN project [DOT14] which is the successor of the highly successful DOTFIVE project [DOT11]. Since such operating frequencies allow circuit operation at 100 GHz, applications such as terahertz imaging and sensing, high-speed communications and the development of measurement equipment for high frequencies ([Sie02,SWH$^+$11]) become possible employing SiGe HBTs. By further aggressive vertical and lateral scaling, transistors with operating frequencies of 1 THz (f_T) and even 1.5 THz (f_{max}) are expected ([SWH$^+$11,SKR$^+$11]), allowing to close the so-called *terahertz gap*. Conventional integrated electronics suffers from a power reduction with increasing operating frequency, making applications at very high frequencies impossible. Optical components, however, show the opposite behavior, leading to a frequency range approximately from 100 GHz to 2 THz[1], where applications are strongly constrained by available devices and concepts. Due to the promising physical limits of SiGe HBTs of more than one THz, they are most likely able to reduce this gap by roughly one decade.

Although emerging technologies such as carbon nanotube field-effect transistors (CNTFETs) also have a theoretical operation speed of several THz ([HSVA06]), existing hardware hardly exceeds 10 GHz ([SCS$^+$12]). Technological and modeling challenges are still too high today to enable circuit operation at similar operating frequencies as by using SiGe HBTs. III-V bipolar technologies, exemplary discussed here for InP HBTs offer similar and even improved operating speed [RLB08] with

---

[1]Different definitions for the limits of the terahertz gap can be found in literature. Here, the most relevant for electronic circuits is used.

also high-speed circuit performance (e.g. [RSS+08]) reported. However, in contrast to III-V devices, integration of SiGe HBTs into CMOS processes is fairly simple, leading to so-called BiCMOS technologies. However, also integration of InP HBTs and Si CMOS on silicon wafers was published (e.g. [KLU+10]). In comparison to RF-CMOS, SiGe HBTs in BiCMOS technologies offer the same speed at a much more depreciated technology node. For example, RF-CMOS with $f_{max} = 400\,\mathrm{GHz}$ was reported at the 28 nm node, while SiGe-HBTs offer the same speed at 130 nm. Thus, circuits with the same performance but of much lower costs can be designed. As a consequence, SiGe HBTs will still play an important role for analog design in the course of the next several years in parallel to other technologies.

In order to utilize the features of SiGe HBTs, though, accurate compact models are required during circuit design and also for technology development and predictive modeling. In this work, rather than deriving a completely new compact model for bipolar transistors, the existing standard model HICUM level 2 (HICUM/L2, [Sch05,SC10]) was used as a fundament. The actual version of the model at the beginning of this work (2.24) is one the two compact models for bipolar transistors supported by the Compact Model Council (CMC), together with MEXTRAM. It is world-wide used in PDKs of foundries which manufacture SiGe HBTs and was recently also successfully applied on InP devices ([NSC+13]). However, lateral and vertical scaling as well as the operation at continuously increasing frequencies revealed the requirement of new and improved model formulations and elements in the equivalent circuit (EC), leading to the motivation of this work. Although in the following chapters the presented application of derived model equations is limited to a single technology, in most cases the model has been applied to a large set of technologies [ARS+10,PSK+11,PS14a]. The equations were also the largest contributor to new model formulations in the release of HICUM/L2 version 2.30 which marked an important milestone in the history of the compact model. Subsequently, it was applied to a number of advanced technologies (e.g. [BSA+12,VDA+12]). Furthermore, the model library extracted in this work has become part of the official PDK for the later described process.

Requirements for the development of new compact model equations are often based on parameter extraction issues for existing technologies. Physical effects not captured by existing compact models often lead to either non-physical parameters or the existing equations are simply not capable of modeling certain characteristics. An example was the extraction of negative c_{10} (or Q_{p0}, respectively, both being parameters of HICUM/L2) for certain technologies when employing standard conventional extraction methods. This effect led to the necessity of an improved model of the transfer current, derived in this work. The important step is to employ accurate TCAD tools to the technology in order to capture the same behavior from device simulations. This step, however, requires accurate transistor parameters such as doping profiles and vertical and lateral dimensions. After successful completion of this step, internal quantities provide an excellent overview on the physical effects causing the actual behavior. In the example it was found that the graded Germanium profile in the base-emitter SCR was leading to a non-ideal transconductance and finally to the failure of the extraction method. Based on the physical equations employed during numerical device simulation, simplified, yet physics-based compact model equations are derived. Those equations are finally applied to the initial data. A comparison of the extracted parameter values with the theory reflects the physical nature of the equations. However, it is crucial to apply the derived model to a large set of different technologies before including it into circuit simulations. Also, the model needs to be tested over a large bias and frequency range covering all possible operation regions of the device to locate potential numerical issues of the equations.

In this work, several aspects of compact modeling are discussed, including simulation of integrated devices, the derivation of physics-based compact model equations, development of extraction routines as well as parameter extraction for fabricated devices. In addition to the discussion of specific effects relevant for SiGe HBTs, it gives a comprehensive overview about compact modeling as a whole.

2 | Modeling of Silicon-Germanium HBTs

2.1 Preface

In this chapter several aspects of modeling SiGe HBTs are discussed. Although the main section (section 2.4) focuses on the effects caused by the introduction of Germanium into the base of a bipolar transistor and is therefore related to SiGe, the remaining observations and models can also be applied to Silicon based transistors. This chapter relies on the basic understanding of the HICUM/L2 compact model and knowledge on the meaning of specific model parameters. The reader is referred to [SC10,SPM13] for a detailed discussion of basic equations and corresponding parameters.

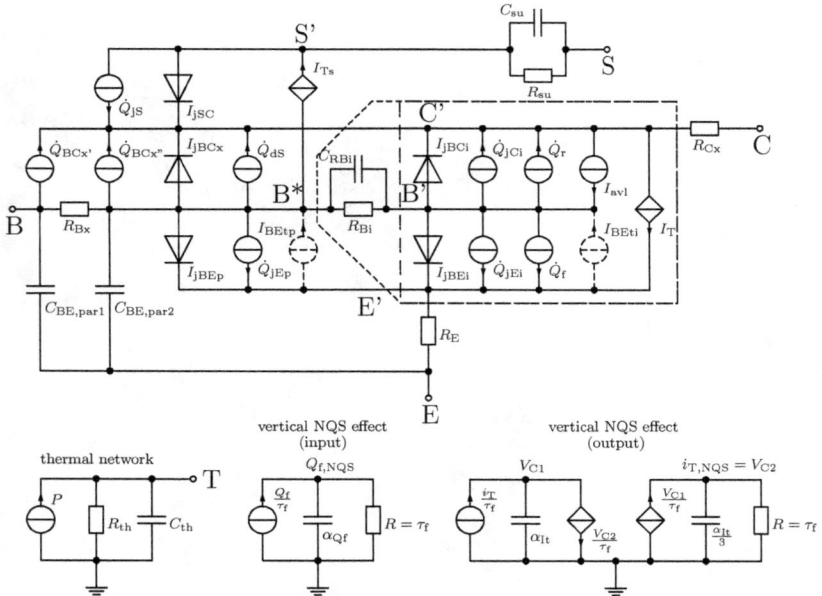

Figure 2.1: Equivalent circuit of HICUM/L2. Shown is the transistor as well as the adjunct networks for modeling the electro-thermal effects (cf. chapter 3.2) and vertical NQS effect (cf. chapter 2.3.2.1). Given by the line with long dashes is the 1D transistor while the internal transistor including the bias and frequency dependent base resistance (cf. 2.3.2.6]) is marked by a line with short dashes.

The equivalent circuit of HICUM/L2 is given in Fig. 2.1. Shown there are all components, adjunct networks and nodes (terminals as well as internal nodes) relevant for this work. Not included are the adjunct networks for the modeling of the noise correlation as well as the noise sources. Most of this chapter discusses effects modeled by elements of the 1D transistor. The external elements relevant in this work are the substrate coupling components (cf. chapter 2.6).

In the beginning of this chapter the existing and widely used numerical simulation approaches for SiGe HBTs are discussed. The subsequent modeling topics comprise the evaluation of the linear and non-linear modeling capabilities of HICUM/L2, namely the modeling of the quasi-static transfer current, the current dependent electric field in the collector and a brief discussion on substrate coupling effects. Without loss of generality in the entire chapter only npn transistors are discussed. However, most of the model equations can also be applied to pnp transistors.

2.2 Numerical simulation of SiGe HBTs

The development of compact models is very often based on numerical device simulations. The latter enable an inside view on the physical processes in semiconductor
devices as they not only provide terminal characteristics but also internal quantities, e.g., carrier densities and velocities. For silicon and silicon-germanium based
bipolar transistors, the transport model employed for numerical simulations is either
a solution of the Boltzmann-transport-equation (BTE) or simplified versions of it,
e.g., the hydro-dynamic (HD) and drift-diffusion (DD) approach. While solving the
BTE is widely regarded as the most physics-based of the three approaches it suffers
from large time (monte-carlo) or memory requirements (direct solution) for running
a simulation. DD presents the classical approach utilized over the last decades for

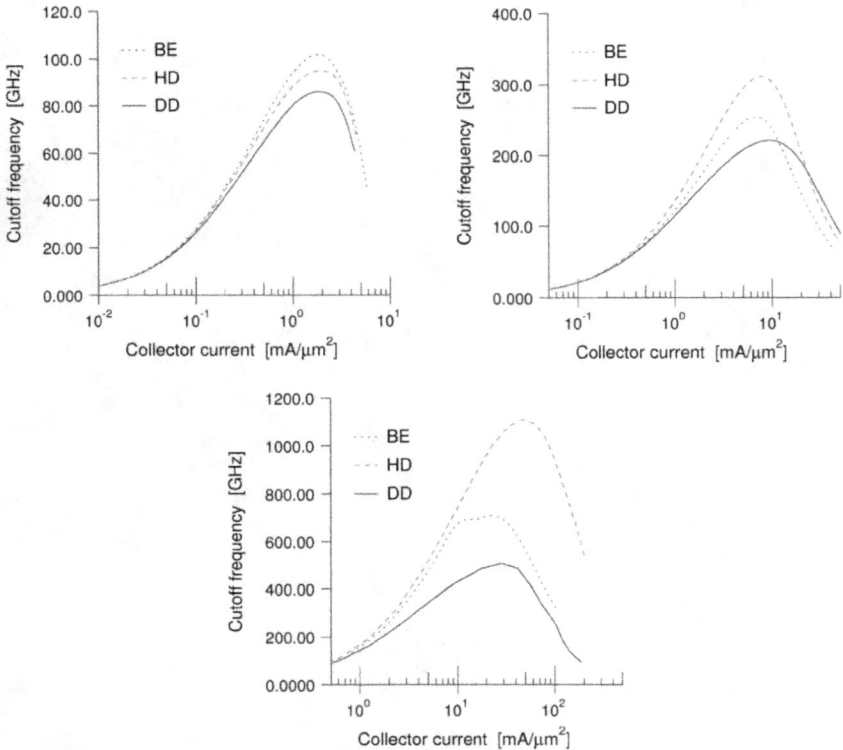

Figure 2.2: Cutoff frequency of different vertically scaled SiGe HBTs simulated by
the BTE, DD and HD approach, adopted from [HJ09].

rapid numerical simulations. However, ongoing vertical device scaling introduces so-called non-local effects, i.e. the carrier velocity is not locally related to the electric field. The HD approach takes the latter effects partially into account by introducing carrier temperatures. A comparison of the different approaches is provided in e.g. [Paw08,HJ09,WS10]. As an example, Fig. 2.2 compares the cutoff frequency of bipolar transistors with different vertical scaling for the three simulation methods. Different results are obtained for the same structure, with DD underestimating and HD overestimating the solution of the BTE. The deviations of the simplified approaches to the reference from the BTE are increasing with increased vertical scaling and, thus, increasing operating frequencies of the device.

Therefore, it is questionable if either of the simplified approaches is capable of describing the actual physical mechanisms in advanced SiGe HBTs correctly. A method to improve results obtained from DD or HD is to calibrate their physical model parameters to the solution of the BTE ([SWH+11]) or to experimental results ([WS10]). However, the latter requires accurate parameters for the external elements of the measured device, because numerical simulations are often carried out for only a spatially limited structure or the vertical 1D profile. After carefully calibrating HD parameters to the BTE, similar results for important FOMs of the HBT can be obtained as presented in Fig. 2.3. However, this calibration is required for each new process node ([SRVC14]). Insufficiently calibrated physical models can lead to non-physical effects, as demonstrated for the output curves in Fig. 2.3. Shown there is the calibration of the HD model by the parameter f^{ec} which modifies the impact of the carrier transport on the energy transport ([WS10]).

The discussion of the general linear and non-linear modeling capabilities presented in this chapter is based on numerical 1D (and partially 2D) device simulations. In both cases only the drift-diffusion equations are solved due to the fact that mixed-mode simulation capabilities, which are required for the non-linear evaluation, are not available for the HD or BTE approach in this work. However, consistent results are required to draw conclusions. In order to reduce the errors due to the simplified transport model a device with moderate speed (peak $f_T = 250\,\mathrm{GHz}$) is used. Although it is shown in [JNM01,PMPK02] that even for slower devices differences between DD and BTE simulations exist and that generally HD can be better calibrated to the BTE (e.g. [HJ08,SWH+11]), results in [HJ09] still show an acceptable match for devices with comparable speed. Furthermore, the general linear modeling capabilities of HICUM/L2 for quasi-static HD simulations as well as a comparison between transit time curves obtained from DD and HD is provided in [Paw08] for current SiGe HBT process generations.

For DD simulations only the time-dependent term of the continuity equation is included. Additional time-dependent terms from the BTE are neglected ([JM05]). However, as shown in later publications ([Jun14]) differences in the small-signal parameters occur only at frequencies higher than $1\,\mathrm{THz}$.

Beyond the general linear and non-linear modeling three effects of Si-BJTs and SiGe-HBTs are discussed in detail. Among them, the substrate coupling can be seen as independent of non-local effects. Therefore, DD simulations are sufficient for providing reference simulations, at least until the onset of frequency dependent

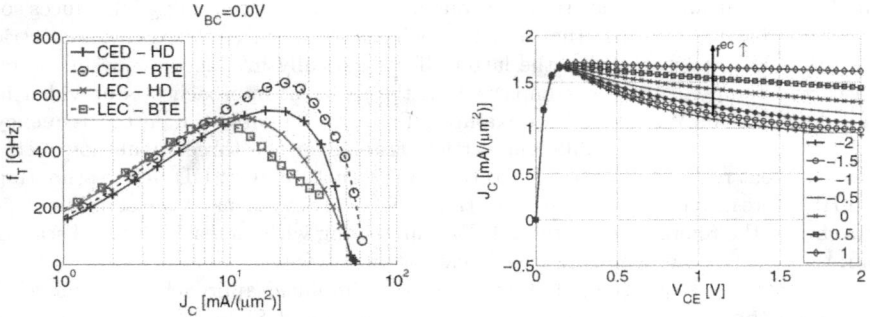

Figure 2.3: HD cutoff frequency of a high-speed SiGe HBT after calibration to BTE and output curves from HD with different parameters, adopted from [WS10].

effects as discussed before. The discussion of the quasi-static modeling is also carried out based on DD results. This has three major reasons. First, the injection of electrons into the base was found to be a major source for deviations of the actual compact model with respect to experimental data. However, this region of the transistor is much less effected by non-local effects as for instance the BC-SCR and the collector. Also, some characteristics of HBTs (e.g. output curves) show non-physical behavior when not carefully calibrated. Second, the modeling is based on the GICCR which in turn is derived from the DD transport equation, due to using the existing compact model HICUM/L2. However, a comparison of the derived model equations to weight factors extracted from HD simulations is presented at the end of the section. Finally, the extracted values as well as the bias and temperature dependences of the weight factors from actual measurements of the latest generation SiGe-HBTs are very close to the values obtained from DD-TCAD simulations.

As mentioned before, non-local effects are mainly located in the base-collector region of the HBT, because the large electric field causes electrons to acquire high energies. However, since the derived model formulation in the corresponding section deviates even to DD simulations in some cases, HD simulations are not performed there.

2.3 Linear and non-linear modeling of bipolar transistors

The aim of this chapter is to provide an overview on the linear and non-linear modeling capabilities of the recent HICUM/L2 version for high-speed applications. Linear operation here is defined by the small-signal parameters, while non-linear modeling is investigated in the context of a simple PA-circuit. The latter is driven with large input power, resulting in harmonic distortion.

2.3.1 Transistor

The transistor utilized in this investigation is displayed in Fig. 2.4. The doping profile and Germanium mole fraction were chosen according to the discussion at the beginning of the chapter. The operating speed of the transistor is given in Fig. 2.5. The maximum frequency of approximately 250 GHz corresponds to the limit of the DD simulation approach.

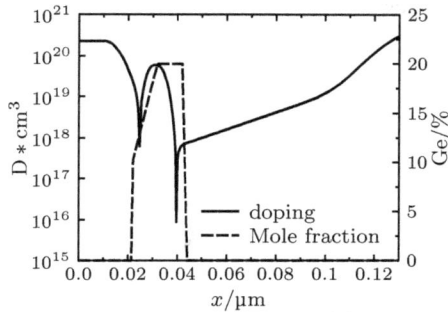

Figure 2.4: Net doping concentration D and Germanium mole fraction for the utilized transistor.

The parameter extraction is performed completely on quasi-static results, i.e. small-signal characteristics like depletion capacitances and f_T are extracted from $d(n, p, \rho)/dV$. The model in Fig. 2.5 shows accurate results for the two most impor-tant FOMs. It is important to note that the I_C model already includes all model equations derived in section 2.4, because this chapter focuses on dynamic modeling. Thus, issues caused by inaccurate modeling of DC quantities are avoided whenever possible.

A zoom into the high current region in Fig. 2.6 shows a good agreement except for very small $V_{C'E'}$. For these bias points, also deviations in f_T exist. It is notable here that a very large value for $g_{\tau fE}$ was required for a reasonable fit of f_T (cf. [Paw14]). This results in a strong overshoot of τ_f as indicated in Fig. 2.6.

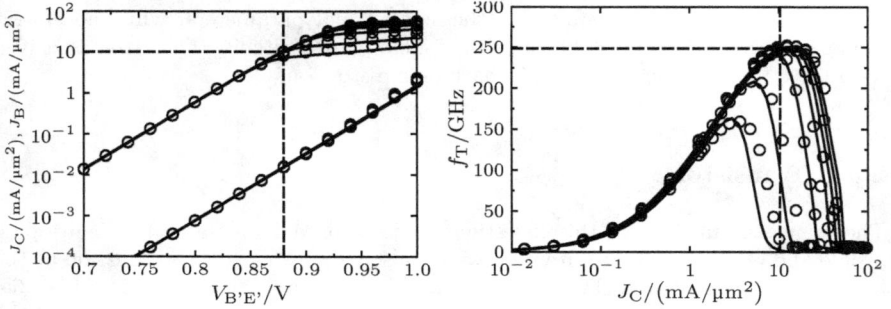

Figure 2.5: Simulated 1D data (circles) and model (lines) of J_C and J_B versus $V_{B'E'}$ and f_T versus J_C. The $V_{C'E'}$ range is from $0.1\,V$ to $2.5\,V$. The operating point according to 2.3.3 is indicated by dashed lines.

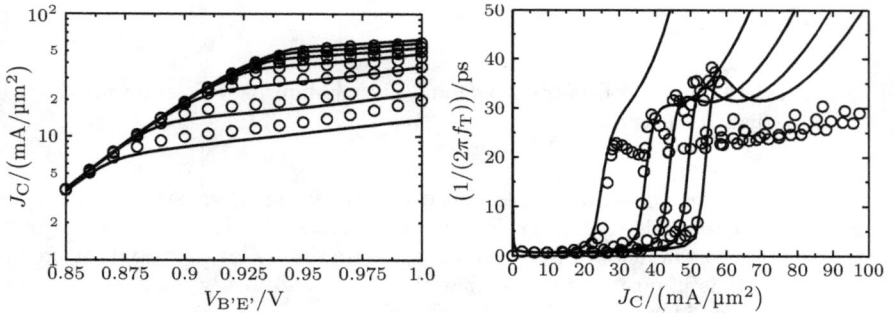

Figure 2.6: Forward gummel characteristics for the same operating points as in Fig. 2.5 and $1/(2\pi f_T)$ for $V_{C'E'} = [0.5\ldots2.5]\,V$.

2.3.2 Linear modeling

The discussion of the linear modeling is based on the small-signal Y-parameters. An overview of the capabilities of the HICUM/L2 model will be presented, starting with the NQS effects and a short evaluation of their implications.

2.3.2.1 Vertical NQS effects

The vertical NQS effects are implemented as explained in [SC10,JDSC10]. Both the delay of the minority charge in the base and the delay of the transfer current are calculated by gyrator networks. The equation for the former is

$$\frac{d\left(\alpha_{Qf}\tau_f Q_{nB}\right)}{dt} + Q_{nB} = Q_{nB0}, \tag{2.1}$$

which is transformed into frequency domain by

$$\underline{C}_{dEi} = \frac{C_{DEi0}}{1 + i\omega\alpha_{Qf}\tau_f}. \tag{2.2}$$

In these equations, Q_{nB} is the actual hole charge and Q_{nB0} its low-frequency (or quasi-static) value. The diffusion capacitances C_{dEi} and C_{DEi0} are similarly defined values. Based on device simulations, the low frequency value C_{DEi0} can easily be extracted from small-signal parameters. Thus, the effective $\underline{C}_{dEi}$[1] follows from

$$\underline{C}_{dEi} = \frac{\underline{Y}_{11} - \left(g_{BE} + i\omega\left(C_{jEi} + C_{jCi}\right)\right)}{i\omega}. \tag{2.3}$$

Note that above value is complex. The extraction of the parameter α_{Qf} is based on (2.2) and given by the equation

$$\alpha_{Qf} = \frac{\Im\left\{C_{DEi0}/\underline{C}_{dEi}\right\}}{\omega\tau_f}. \tag{2.4}$$

The equation for the delay of the transfer current reads

$$\frac{d^2(\alpha_{IT}^2/3)\tau_f^2 i_{Tf}}{dt^2} + \frac{d\alpha_{IT}\tau_f i_{Tf}}{dt} + i_{Tf} = i_{Tf0} \tag{2.5}$$

and reads in frequency domain

$$\underline{g}_m = \frac{g_{m0}}{1 + i\omega\alpha_{IT}\tau_f - \omega^2\alpha_{IT}^2/3\tau_f^2} = g_{m0}\frac{1 - \omega^2\alpha_{IT}^2/3\tau_f^2 - i\omega\alpha_{IT}\tau_f}{\left(1 - \omega^2\alpha_{IT}^2/3\tau_f^2\right)^2 + \left(\omega\alpha_{IT}\tau_f\right)^2}. \tag{2.6}$$

From this equations follows an imaginary part for $\underline{g}_m$ even at low frequencies. For operating points with a large g_{m0} and τ_f this part can exceed ωC_{jCi}. Thus, it is not possible to extract C_{BC} correctly from $\Im\{\underline{Y}_{21}\}$ in those regions. The parameter

---

[1]The value is labeled as "effective" because the terminal Y-parameters are used as reference rather than internal quantities.

α_{IT} is extracted from (2.6) by using either the real or imaginary part of the term $g_{\mathrm{m0}}/\underline{g}_{\mathrm{m}} - 1$.

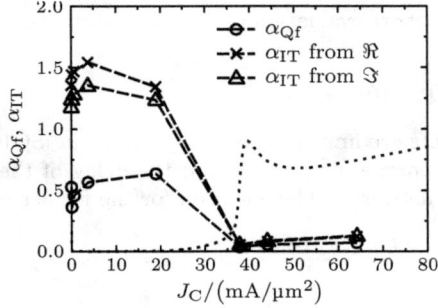

Figure 2.7: Extracted values for α_{Qf} and α_{IT} for different operating points. "α_{IT} from $\Re$" means the value is extracted from the real part of (2.6). "α_{IT} from $\Im$" is similarly defined. The dotted line shows the τ_{f} curve.

Extraction results for both parameters are shown in Fig. 2.7. In the medium current region non-physical large values are extracted for both[1], although the ratio of both is in the expected range. In the high current region the parameters drop to physical values. This observation was confirmed by [Hus]. Possible explanations for these effects are missing series resistances[2] or the phase shift in the BC-SCR. In the latter case, the continuity equation simplifies for small-signal operation at low and medium bias to

$$\frac{\mathrm{d}\underline{J}_{\mathrm{n}}}{\mathrm{d}x} = \mathrm{i}\omega\frac{-\underline{J}_{\mathrm{n}}}{v_{\mathrm{sn}}}, \qquad (2.7)$$

with the solution

$$\underline{J}_{\mathrm{n}}(x) = \underline{J}_{\mathrm{n,xc}} \exp\left(-\mathrm{i}\omega\frac{x - x_{\mathrm{c}}}{v_{\mathrm{sn}}}\right). \qquad (2.8)$$

Here, $\underline{J}_{\mathrm{n,xc}}$ is the small-signal current density at the start of the BC-SCR. A linear phase shift of the collector current is obtained, which is not included in the current formulation of the vertical NQS effect. However, this issue will not be discussed in further detail. Also, recently published approaches as in, e.g., [HC14] have to be evaluated in the future.

Results for the corresponding Y-parameters are given in section 2.3.2.2 and 2.3.2.3. Following from the actual implementation of the model by adjunct networks, the NQS effects are calculated based on the change of I_{Tf} and Q_{dEi}. However, at high currents, both become dependent also on $V_{\mathrm{C'E'}}$ due to $I_{\mathrm{CK}}(V_{\mathrm{C'E'}})$. Thus, also the derivatives of both quantities with respect to $V_{\mathrm{B'E'}}$ and $V_{\mathrm{C'E'}}$ are based on the

---

[1] Physical values for the drift transistors are $\alpha_{\mathrm{Qf}} \approx 0.2$ and $\alpha_{\mathrm{IT}} \approx 0.5$.

[2] Series R_{E} and R_{Cx} exist in 1D device simulations due to the neutral regions. However, their impact on the Y-parameters was evaluated by deembedding procedures and found to be negligible ([Paw14]).

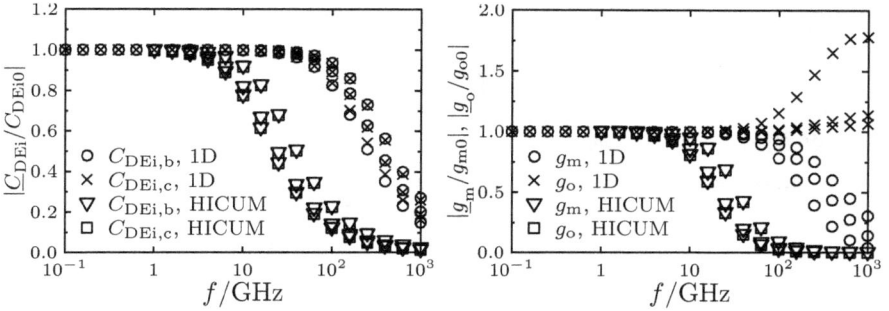

Figure 2.8: Frequency dependence of the normalized diffusion capacitance from $\underline{Y}_{11}$ and $\underline{Y}_{12}$ as well as $\underline{g}_{\mathrm{m}}$ and $\underline{g}_{\mathrm{o}}$. Shown is the comparison between 1D numerical simulations and HICUM simulations. Note that the operating points for the 1D simulation and HICUM are not comparable. Large differences between both are intentional to improve the clarity of the pictures.

same equations. This is shown in Fig. 2.8. Here, the different diffusion capacitances are defined as

$$C_{\mathrm{DEi0,b}} = \frac{\mathrm{d}Q_{\mathrm{dEi}}}{\mathrm{d}V_{\mathrm{B'E'}}} \quad \text{and} \quad C_{\mathrm{DEi0,c}} = \frac{\mathrm{d}Q_{\mathrm{dEi}}}{\mathrm{d}V_{\mathrm{C'E'}}}. \tag{2.9}$$

The frequency dependences of $\underline{C}_{\mathrm{DEi,b}}$ and $\underline{C}_{\mathrm{DEi,c}}$ are identical in HICUM/L2, as are the ones for the frequency dependent transconductance $\underline{g}_{\mathrm{m}}$ and output conductance $\underline{g}_{\mathrm{o}}$. 1D device simulations show in contrast that this effect holds only for the diffusion capacitance. The frequency dependence of $\underline{g}_{\mathrm{o}}$ shows the opposite behavior of $\underline{g}_{\mathrm{m}}$. Thus, errors in the modeling of $\underline{g}_{\mathrm{o}}$ for high frequencies are inevitable for the 1D transistor. However (at high frequencies) in real devices the substrate-effects dominate the frequency dependence $\underline{Y}_{22}$ for of both the real and imaginary part. The errors due to the NQS modeling are therefore less pronounced for real devices, yet might be relevant for SOI devices.

2.3.2.2 Y_{11}

The implications following from the strongly differing values of α_{Qf} at medium and high injection presented in Fig. 2.7 are highlighted in Fig. 2.9. Compact model simulations were performed utilizing a large α_{Qf} obtained for operating points in the medium-current region and with a small α_{Qf} extracted in the high-current region. Obviously, not all operating regions can be modeled accurately by a single α_{Qf} value. However, in order to avoid an overestimation of the NQS effects, the smaller value is used for the model employed in 2.3.3.

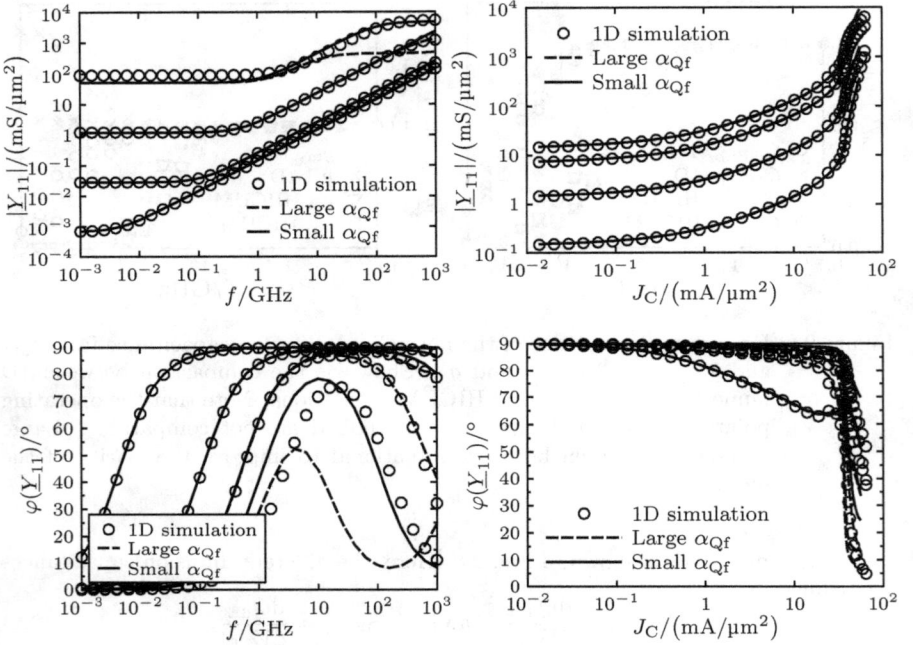

Figure 2.9: Magnitude and phase of $\underline{Y}_{11}$ as a function of frequency for collector current densities $J_C = [0.01, 0.6, 20, 50]\,\mathrm{mA/\mu m^2}$ and as a function of operating point for $f = [1, 10, 50, 100]\,\mathrm{GHz}$. $V_{B'C'} = 0\,\mathrm{V}$ is used in both cases. Here, "large α_{Qf}" means an average value in the medium current region, i.e. before the onset of high current effects, while "small α_{Qf}" is a value extracted in the high-current region (cf. Fig. 2.7).

2.3.2.3 Y$_{21}$

For $\underline{Y}_{21}$, the same conclusions as for $\underline{Y}_{11}$ in 2.3.2.2 can be drawn. The results are shown in Fig. 2.10. As shown in Fig. 2.11, a large α_{IT} results in an improved model for the imaginary part of $\underline{Y}_{21}$. In the quasi-static part of the HICUM/L2 model, before the onset of high-current effects, $\Im\{\underline{Y}_{21}\}$ is only defined by C_{jCi} and should yield a constant value versus I_C (for $V_{B'C'} = 0\,\mathrm{V}$). The observed increasing $\Im\{\underline{Y}_{21}\}$ can be caused by two effects. First, already a small phase shift due to NQS effects has significant influence on this value, while it has almost none on the much larger real part. This is in fact what is happening in the model. Second, the current dependence of the BC-depletion capacitance causes an increase of the collector charge, leading to an increase in the imaginary part of $\underline{Y}_{21}$, too. This effect is discussed in detail in 2.5. Summarized, except for the issues related to the missing phase shift of the collector current in the BC-junction also the current dependence of C_{jCi} can cause too large extracted values of α_{IT}.

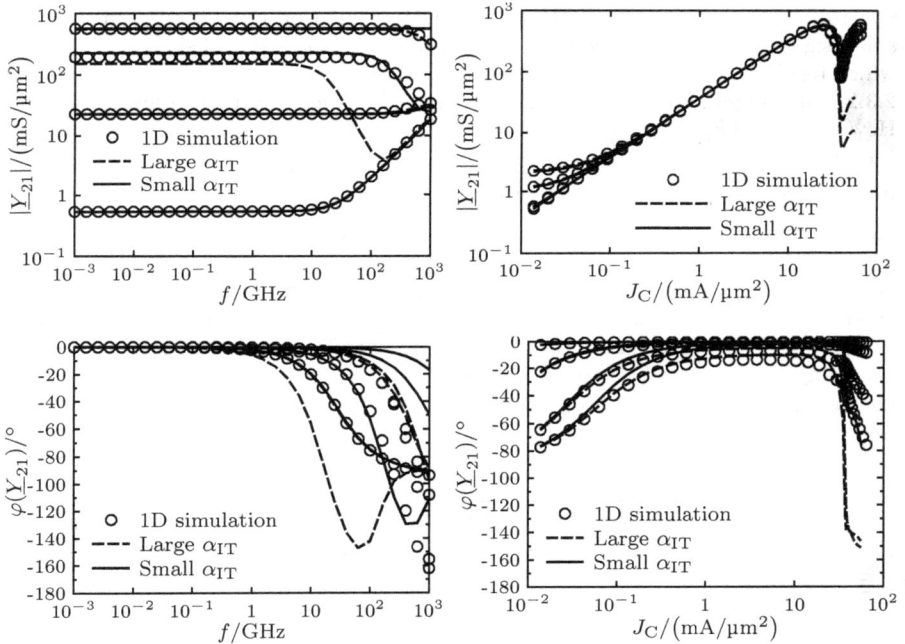

Figure 2.10: Magnitude and phase of $\underline{Y}_{21}$ for the same conditions as in Fig. 2.9. The meaning of "large α_{IT}" and "small α_{IT}" is similar to that of Fig. 2.9.

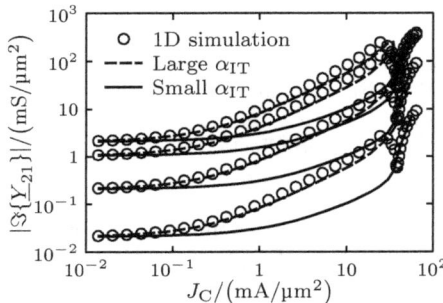

Figure 2.11: Imaginary part of $\underline{Y}_{21}$ as a function of bias for $f = [1, 10, 50, 100]\,\mathrm{GHz}$.

2.3.2.4 Y_{12}

While comparing the model of $\underline{Y}_{12}$ to the reference data from 1D simulations, different effects are observed. An issue pronounced in advanced SiGe technologies is the neutral base recombination (NBR). The effect of NBR on the base current and the small-signal parameters is shown in Fig. 2.12. As one can see I_B decreases with $V_{C'E'}$ which results in a further decrease of the negative real part of $\underline{Y}_{12}$. As explained in 2.3.2.1 and highlighted in Fig. 2.12(b), NQS effects also have an influence of $\underline{Y}_{12}$. However, the shape of $\Re\{\underline{Y}_{12}\}$ cannot be explained by NQS effects.

Although a physics-based model was recently proposed in [Sch14], during this investigation an empirical model with

$$I_{BEis} = I_{BEis,0} \left(1 + \alpha_{IBEs} V_{B'C'}\right) \tag{2.10}$$

was employed. Using this simple model, accurate results for I_B are obtained. Yet the model is not suitable for production compact models, because the base current would reverse for (extremely) small $V_{B'C'}$, it is sufficient for the discussion here. Including this model fixes the wrong model for $\underline{Y}_{12}$ at low bias. However, NQS effects still negatively influence the real part at medium and high current densities.

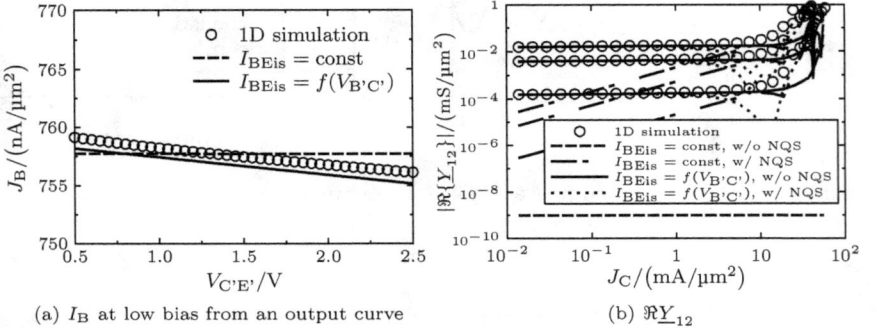

(a) I_B at low bias from an output curve

(b) $\Re\underline{Y}_{12}$

Figure 2.12: (a) Base current for an output curve at low bias with and without the model (2.10). (b) Real part of $\underline{Y}_{12}$ versus collector current for the same conditions as in Fig. 2.9 (except for $f = 1\,\mathrm{GHz}$) with and without the model (2.10) and NQS effects.

In addition, issues due to the barrier model were found for $\Im\{\underline{Y}_{12}\}$. As shown in Fig. 2.13, a change of the sign of $\Im\{\underline{Y}_{12}\}$ occurs in the actual implementation which is not observed in the reference 1D simulations. This effect also occurs at low frequencies and is therefore not correlated to NQS effects.

For forward operation (i.e. $C_{dCi} \approx 0$) the imaginary part of $\underline{Y}_{12}$ for the 1D case

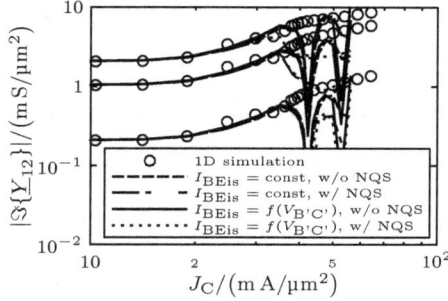

Figure 2.13: Imaginary part of $\underline{Y}_{12}$ versus collector current for $f = [10, 50, 100]$ GHz. The different compact model results are for the same models as in Fig. 2.12.

at low frequencies and without series resistances is given by

$$\Im\{\underline{Y}_{12}\} = 2\pi f \left(C_{jCi} + \left. \frac{dQ_{dEi}}{dV_{C'E'}} \right|_{V_{B'E'}} \right). \tag{2.11}$$

The latter component of above equation is caused by the $V_{B'C'}$- and $V_{C'E'}$-dependence of Q_{dEi} and the resulting dynamic current. It is given by

$$\left. \frac{dQ_{dEi}}{dV_{C'E'}} \right|_{V_{B'E'}} = \left. \frac{dQ_{dEi}}{dV_{C'E'}} \right|_{V_{B'E'}} \tag{2.12}$$

$$= -\frac{d\tau_{f0}}{dV_{B'C'}} I_{Tf} + \left. \frac{dQ_{fh}}{dI_{CK}} \right|_{I_{Tf}} \frac{dI_{CK}}{dV_{C'E'}} + (\tau_{f0} + \Delta\tau_{fh}) \left. \frac{dI_{Tf}}{dV_{C'E'}} \right|_{V_{B'E'}},$$

where the last two terms (without τ_{f0}) dominate in the high current region. The value of the imaginary part of $\underline{Y}_{12}$ is a compensation of two effects. With increasing I_{CK} at a fixed I_{Tf}, the charge is decreasing. On the other hand, the charge increases due to the increase of I_{Tf} for changing $V_{C'E'}$. Both parts are investigated further next, starting with the latter component.

This component can be calculated directly from the low frequency Y-parameters using

$$(\tau_{f0} + \Delta\tau_{fh}) \frac{dI_{Tf}}{dV_{C'E'}} = \left(\frac{\Im\{\underline{Y}_{11}\}}{2\pi f} - C_{jEi} - C_{jCi} \right) \frac{\Re\{\underline{Y}_{22}\}}{\Re\{\underline{Y}_{21}\}}. \tag{2.13}$$

A small error is made due to neglecting the $d\tau_{f0}/dV_{B'C'}$ related term. However, although this term is multiplied with I_{Tf}, the resulting error is very small. The term related to $dQ_{fh}/dI_{CK}|_{I_{Tf}}$ cannot be extracted directly from device simulations. Therefore it is calculated by (cf. (2.12))

$$\left. \frac{dQ_{fh}}{dI_{CK}} \right|_{I_{Tf}} \frac{dI_{CK}}{dV_{C'E'}} = \left(\frac{\Im\{\underline{Y}_{12}\}}{2\pi f} + C_{jCi} \right) - (\tau_{f0} + \Delta\tau_{fh}) \frac{dI_{Tf}}{dV_{C'E'}}, \tag{2.14}$$

where again the $d\tau_{f0}/dV_{B'C'}$ related term is neglected. Results for both components are given in Fig. 2.14.

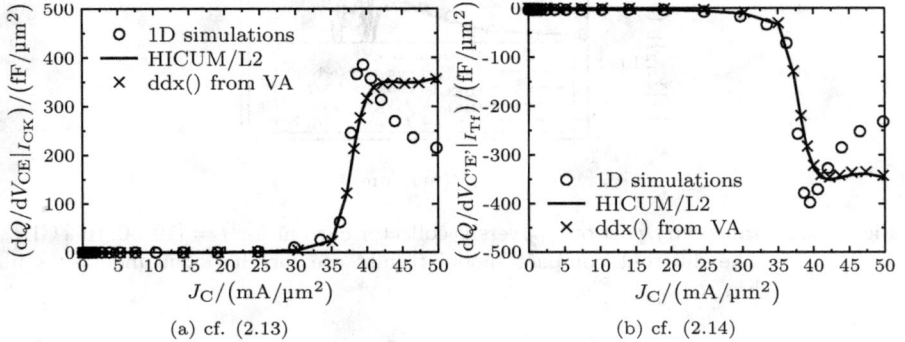

(a) cf. (2.13)   (b) cf. (2.14)

Figure 2.14: Derivatives of the minority charges with respect to constant I_{CK} and I_{IT}. Here, the "1D simulations" and "HICUM/L2" represent results directly extracted from simulated Y-parameters. "ddx() from VA" is the value calculated in the VA code to show the impact of the neglected quantities.

Since both components have a similar shape with opposite sign, the actual sign of $\Im\{\underline{Y}_{12}\}$ depends on the small difference between the absolute values of both. As shown in Fig. 2.15(a), the impact of both components from τ_{f0}, i.e. $-d\tau_{f0}/dV_{B'C'}I_{Tf}$ and $\tau_{f0}g_0$, cannot be completely neglected but does not affect the sign change. Therefore, it is sufficient to evaluate only the high current minority transit times further. A summary of the derivatives of the diffusion charge is given in A.1. From these equations follows that the increase (and thus the change of the sign) is caused by the barrier charge $Q_{Bf,b}$, shown in Fig. 2.15(b).

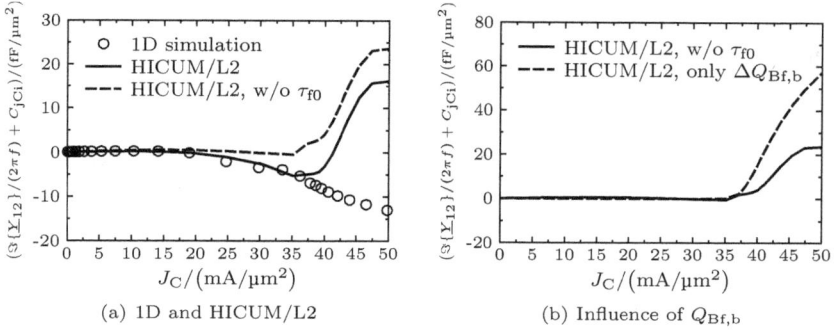

(a) 1D and HICUM/L2

(b) Influence of $Q_{Bf,b}$

Figure 2.15: (a) Modeling of $\Im\{\underline{Y}_{12}\}$ without the impact of the components caused by τ_{f0}. (b) Derivative of the barrier charge $Q_{Bf,b}$.

2.3.2.5 Y_{22}

The final results for $\underline{Y}_{22}$ are given in Fig. 2.16. In Fig. 2.16(a) the influence of the NQS effects on $\underline{Y}_{22}$ as discussed in 2.3.2.1 is visible. Although the frequency dependence at low bias is directly correlated to the series R_{Cx}, the actual NQS model leads to wrong results at very large bias.

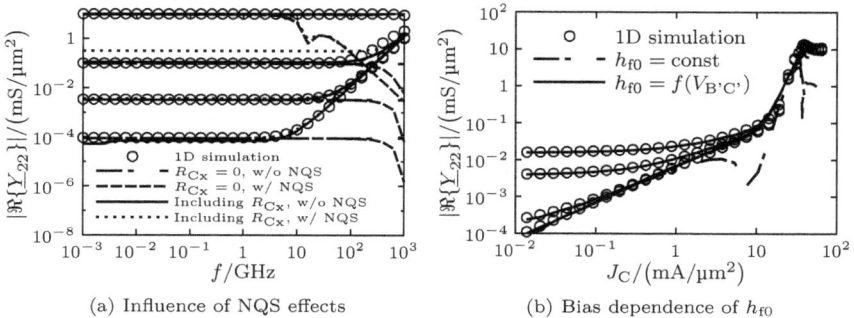

(a) Influence of NQS effects

(b) Bias dependence of h_{f0}

Figure 2.16: (a) Real part of $\underline{Y}_{22}$ versus frequency for the same bias as in Fig. 2.9. (b) Magnitude of $\underline{Y}_{22}$ versus collector current density.

Fig. 2.16(b) shows a completely wrong bias dependence of the real part at medium bias, independent of the frequency. This behavior is discussed in detail in 2.4.3.2, where a model for the weight factor h_{f0}[1] as a function of bias (especially on $V_{B'C'}$)

---

[1]The weight factor h_{f0} is not available in older HICUM/L2 versions but is introduced in this work.

is developed. Very low values for the output conductance and even negative values are caused by $d\tau_{f0}/dV_{B'C'}$ and are eliminated by a correct modeling of the bias dependence of h_{f0}.

2.3.2.6 Dynamic emitter current crowding

Note that in this section currents rather than current densities are used since the scaling of R_{Bi} is different compared to currents.

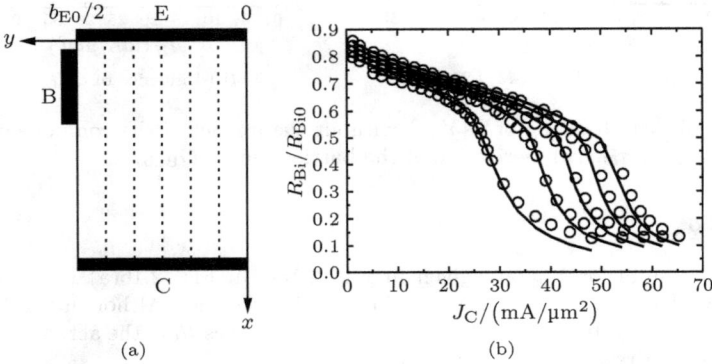

Figure 2.17: (a) Sketch of the ideal 2D structure. The dashed lines represent the discretization in lateral dimension (y). (b) Model (solid lines) of R_{Bi} versus the extracted data (circles) based on (2.15) for $V_{B*C'} = [-0.5 \ldots 1.5]\,\text{V}$. R_{Bi} is normalized to its zero-bias value R_{Bi0}.

So far only 1D numerical simulations were discussed. For the sake of completeness a short overview regarding the extensions to 2D simulations covering the complete internal transistor including R_{Bi} is presented in this section. The corresponding numerical simulations were performed with an added lateral discretization below the emitter. No external doping profile was added. The standard 1D simulation contains only a single lateral discretization point or two. For the latter the same bias is applied to avoid lateral current flow. By adding additional lateral points below the emitter this lateral current flow is enabled resulting in static and dynamic current crowding. A sketch of an ideal 2D structure with the emitter width b_{E0} is shown in Fig. 2.17 (a). Note that at 0 a symmetry line is added. The extraction of R_{Bi} from numerical simulations can be performed according to

$$R_{Bi} = \frac{V_{B'E'} - V_{B*E'}}{I_B}, \tag{2.15}$$

where $V_{B'E'}$ and $V_{B*E'}$ are the terminal voltages of the 1D and ideal 2D transistor, respectively. Both are extracted at the same I_B. The same extraction can be performed based on I_C. The model for R_{Bi} as function of bias (for model equations refer to [SC10]) is given in Fig. 2.17 and provides accurate results in the complete investigated operating region.

For very small switching times or very high frequencies dynamic emitter current crowding leads to a reduction of the base impedance. This effect is often referred to as *lateral NQS effect*. Models for both large- and small-signal operation were developed, yet they are either to complicated or simply not suitable for integration into a compact model, because the derived equations cannot be represented by circuit elements ([Pri58,Rei77,SR95]). The most commonly used model [Ver91,SC10] adds a capacitance in parallel to R_{Bi} for modeling the first-order small-signal effects. Results of this model using the extracted HICUM/L2 parameter $f_{CRBi} = 0.3$ are provided in Fig. 2.18. Here, $\underline{Z}_{Bi}$ is the small-signal base impedance which was extracted similarly to (2.15) by

$$\underline{Z}_{Bi} = \frac{1 - \underline{Y}_{11,2D}/\underline{Y}_{11,1D}}{\underline{Y}_{11,2D}}. \tag{2.16}$$

$\underline{Y}_{11,1D}$ and $\underline{Y}_{11,2D}$ are the values from 1D and 2D simulation for the same I_B. Although the magnitude can be modeled with an acceptable agreement, the phase shift is strongly overestimated.

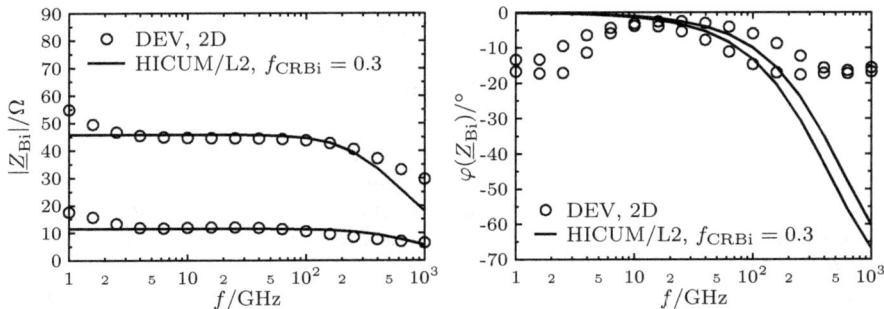

Figure 2.18: Magnitude and phase of the frequency dependent base impedance for $V_{B^*E'} = 0.9\,\text{V}$ and $1.0\,\text{V}$.

Very recently, an approach employing a simplified distributed model (similar to [Rei77]) was published in [SCS14] and provides a good agreement over a large frequency range. However, this model was not taken into account in this work.

2.3.3 Non-linear modeling

The verification of the non-linear modeling capabilities of HICUM/L2 is based on mixed-mode numerical device simulations of a simple power amplifier (PA) and inspired by the work made in [SPL00]. Only harmonic distortion will be discussed, since transient numerical simulations required for the investigation of effects by in-termodulation distortion are far too time consuming, especially when using realistic high frequencies but small offsets between the frequencies. A verification in both frequency- and time-domain results is given.

2.3.3.1 Circuit

The simulated PA-circuit is given in Fig. 2.19. It consists of two DC voltage sources for the bias point and a sinusoidal voltage source V_S with the internal resistance R_S, which represent the power source. Also included are DC feed inductors and blocking capacitors.

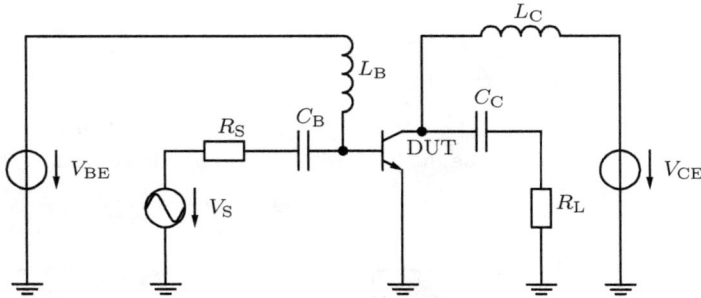

Figure 2.19: Schematic of the simulated circuit of a power amplifier.

During simulation, the source resistance was set to $50\,\Omega$. The load impedance was calculated in order to have an optimum load line under DC conditions. Therefore, only a resistance R_L is defined. Although no optimal operation of the circuit as a PA is obtained by this setup for high frequencies, it is sufficient for an investigation of the non-linear model capabilities of HICUM/L2.

Since only transient simulations (and no steady-state simulations, cf. 4.4.5.2) are possible with the utilized device simulator, values for the feed inductors and blocking capacitors have to be defined[1]. The aim is to have almost ideal behavior by still providing reasonable simulation times. The values used during simulation are related to the load impedance by

$$L_C = \frac{25 R_L}{2\pi f} \quad \text{and} \quad C_C = \frac{1}{2\pi f 5 \times 10^{-3} R_L}. \tag{2.17}$$

2.3.3.2 Operating point

The operating point for this investigation was chosen to be $V_{B'E'} = 0.88\,\text{V}$ and corresponds to a current density of approximately $J_C = 10\,\text{mA}/\mu\text{m}^2$. From the

---

[1] For Harmonic Balance simulations, using $L = C = \infty$ is possible. For transient simulations however, too large values result in too large simulation times before the steady-state is reached.

chosen $V_{C'E'} = 1.0\,\text{V}$ follows $R_L = 100\,\Omega$.

The operating point is given in Fig. 2.5 and the resulting static load line in Fig. 2.20. The chosen operating point is very close to peak f_T and, thus, less relevant for a practical circuit application. However, for the intended evaluation of the modeling capabilities this operating point is well suited, because it also leads to non-linearities caused by high current effects. In [Paw14], the same investigations were performed for more suited operating points in terms of circuit design.

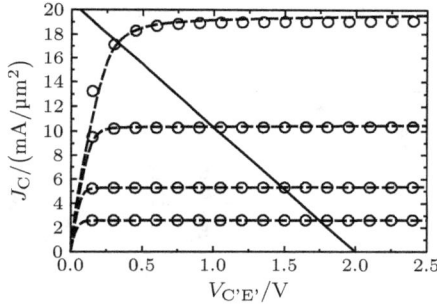

Figure 2.20: 1D numerical simulation (circles) and model results (dashed lines) for output curves at $V_{B'E'} = [0.84\ldots0.9]\,\text{V}$. Included as solid line is the static load line.

2.3.3.3 Time-domain results

The comparison of time domain results shown here is provided for two frequencies. They roughly represent the upper limit of existing equipment for non-linear one-tone measurements $(10\,\text{GHz})$[1] and the upper limit of the operation frequency for the transistor $(200\,\text{GHz})$, resulting in a large impact of NQS-effects. In order to evaluate the influence of the latter, three different models are employed, differing in the parameters for the vertical NQS-effects. They are summarized in Tab. 2.1 (cell headings there correspond to legend entries of subsequent figures). The NQS model parameters are based on the results in Fig. 2.7 but are more relaxed for the "large NQS" case.

Parameter	w/o NQS	small NQS	large NQS
fl_{NQS}	0	1	1
α_{Qf}	0	0.05	0.5
α_{IT}	0	0.05	0.38

Table 2.1: Model cards for non-linear modeling.

---

[1]A sufficient number of measured harmonics, i.e. a minimum of five, is assumed.

For 10 GHz results for the time dependent i_C and i_B are given in Fig. 2.21[1] for a high input power[2]. The corresponding 1 dB compression point is at $P_{avs} = -23$ dBm. Except for very small deviations at maximum i_C both collector and base current are modeled very accurately. The curve for the collector current clearly shows no impact of dynamic currents as was expected for the chosen operating point and frequency, whereas dynamic currents dominate the base current. However, as also expected for the relatively small frequency NQS effects are not affecting both currents. Thus, even for an operating point at peak f_T and very high source power the fundamental frequency of 10 GHz can be declared as quasi-static in terms of vertical NQS effects.

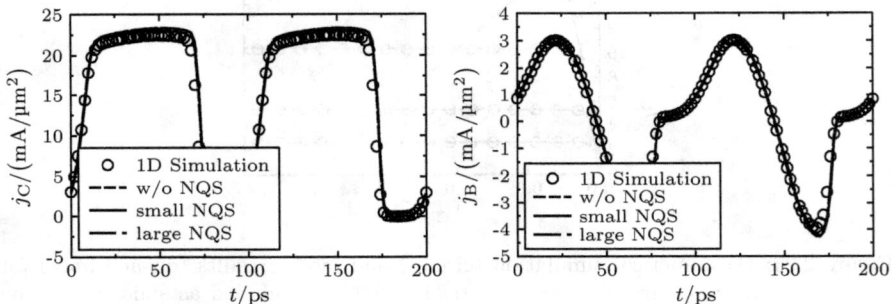

Figure 2.21: Collector and base current at $P_{avs} = -10$ dBm and $f = 10$ GHz. The models used for simulation are defined according to Tab. 2.1.

From the simulation results for 200 GHz presented in Fig. 2.22, where the smaller input power corresponds to $P_{in,1dB}$, two findings are obtained. For relatively small power the employed large NQS parameters provide a more accurate model for $V_{B'E'}$ which is distorted by the voltage drops across R_S. However, with increasing power the delay in i_C is still modeled correctly by the large NQS parameter at the beginning of the rising slope, but a strong overestimation of the delay is modeled starting at approximately the second half of this slope. In contrast, the small NQS parameters lead to a small oscillation. Avoiding the model for the NQS effects completely still provides the most accurate curve shapes at this frequency, although the shift in the delay is missing. A short discussion of the inaccurate model for i_C when including NQS effects in the compact model is provided in the next section.

2.3.3.4 Frequency-domain results

Applying a Fourier series to the terminal voltages and currents allows the calculation of frequency domain quantities, defined by the components for the fundamental

---

[1]In all time domain plots $v_{B'E'}(t)$ is not included for reasons of clarity.

[2]Since the transducer gain of a device decreases with frequency for fixed source and load impedance, results for much higher source power values are shown for 200 GHz.

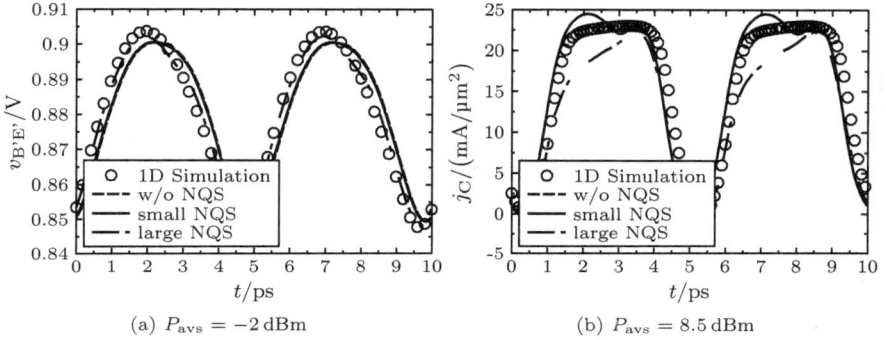

Figure 2.22: Base-emitter voltage and collector current at different source power values for $f_0 = 200\,\text{GHz}$. The models used for simulation are defined according to Tab. 2.1.

frequency f_0 and all harmonic frequencies. Since in this chapter only steady-state simulations are performed, no continuous but a discrete spectrum is obtained. Components are discussed as complex values. The real and imaginary part are calculated from the cos- and the sin-part of the Fourier series.

The comparison in frequency domain is carried out based on practice-oriented quantities. The output power is given by

$$P_{\text{out}} = 0.5\Re\{\underline{V}_{\text{CE}}\overline{\underline{I}_{\text{C}}}\}, \qquad (2.18)$$

and can be defined for the fundamental and each harmonic frequency. Here, $\overline{\underline{I}_{\text{C}}}$ is the complex conjugated of the output current. Important characteristic quantities including but not limited to:

- the $1\,\text{dB}$ compression point $P_{\text{in,1dB}}$, i.e. the input power, where the fundamental output is reduced by $1\,\text{dB}$ compared to the ideal - from low input power extrapolated - value, and

- the intercept point for the third harmonic (also defined from extrapolation based on low input powers) which is an important figure of merit for classifying non-linear systems

can be derived. A comprehensive discussion on the modeling of those quantities is given in [Paw14,MP14].

The transducer gain is based on the maximum available input power from the source, reading

$$G_{\text{T}} = \frac{P_{\text{out}}}{P_{\text{avs}}}, \qquad (2.19)$$

whereas the power added efficiency PAE is calculated based on the actual input

power

$$\text{PAE} = \frac{P_{\text{out}} - P_{\text{in}}}{P_{\text{DC}}}, \tag{2.20}$$

with the input power for the device $P_{\text{in}} = 0.5\Re\{\underline{V}_{\text{BE}}\overline{\underline{I}_{\text{B}}}\}$ with the complex conjugated input current $\overline{\underline{I}_{\text{B}}}$ and the dissipated DC power estimated by $P_{\text{DC}} = V_{\text{BE}}I_{\text{B}} + V_{\text{CE}}I_{\text{C}}$. Both are only calculated from the fundamental components of the corresponding quantities. However, they are also affected by the intermodulation distortion. Furthermore, an important figure of merit for non-linear modeling is the DC bias I_{C} as function of input power. Even harmonics generate a shift in the operating point resulting in an increasing power dissipation by the supply and, thus, have to be modeled correctly.

Results are not discussed here, since all quantities are more or less directly represented by the model of $\underline{I}_{\text{C}}$ which is the only quantity shown here. They are compared to actual measurements in 4.4.5. As shown in Fig. 2.21 the model for the time domain i_{C} is very accurate for low frequencies, directly leading to accurate results in frequency domain. However, especially the model for the NQS effects causes undesired behavior at higher frequencies as highlighted in Fig. 2.22.

Large NQS parameters cause an over-compression of the harmonic components beginning with the second one. Thus, since the fundamental frequency is - at least at the given input power - still modeled accurately, the delay of i_{C} is correct. However, the smaller magnitude of the harmonic components (cf. Fig. 2.23(b)) cannot support the desired fast switching. In contrast, the phase delay is incorrect as shown in Fig. 2.22, although the magnitudes are modeled correctly (cf. Fig. 2.23(a)) employing the small NQS parameters.

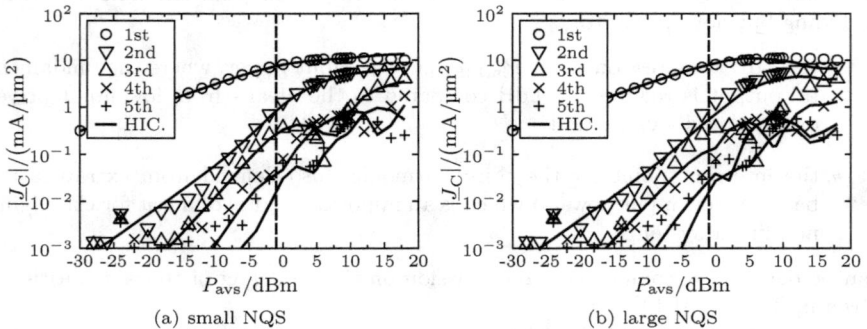

(a) small NQS   (b) large NQS

Figure 2.23: Fundamental and harmonic components of the collector current versus maximum available input power at $f = 200\,\text{GHz}$.

2.3.4 Summary

The overview given in this chapter proves the accurate modeling of SiGe transistors with HICUM/L2 in small- and large-signal operation, but also addresses issues with the models for both the vertical and lateral NQS effect. While no further discussion of the latter was provided, results for the former can be interpreted as follows: The actual model for the vertical NQS effects is not capable of modeling the phase shift and amplitude of the collector current simultaneously. While the phase shift leads to large values for the NQS parameters during extraction, using those results in a strong underestimation of the actual amplitude. Possible reasons for this behavior are the neglected phase shift of the collector current in the BC-SCR and the current dependent capacitance.

Additional error terms were identified in certain model equations, including the barrier term causing the wrong sign of $\underline{Y}_{12}$ and the incorrect influence of the I_C NQS model on $\underline{Y}_{12}$. Nevertheless, overall highly accurate results accentuate the good reputation of the HICUM/L2 model for the modeling of SiGe-HBTs. It is thus the perfect choice for all following investigations.

2.4 Quasi-static modeling of the transfer current

2.4.1 Physical effects in heterojunction bipolar transistors

The production of SiGe-HBTs with a graded Germanium profile in the base in order
to reduce the base transit time led also to an introduction of new physical effects
compared to Si-BJTs. In early publications about the low temperature applications
of SiGe-HBTs, the focus was laid especially on the strong bias dependence of the
current gain. The latter was also found being a strong function of the device temperature ([CPS+90]). An explanation of the effect was later given by the same authors
in [CCP+93]. It was shown that the bias dependence is caused by the shift of the
base-side edge of the base-emitter SCR. For graded germanium profiles, the electrons are injected at different bandgap and, thus, different intrinsic carrier density.
A review of the effect with a detailed discussion follows in Ch. 2.4.1.2. Later, the
impact on temperature sensitive circuits was among others discussed in [SCJH00].
There, the difference between the bandgap in the neutral base and the end of the
BE-SCR was identified as largest contributor to this effect. The first equations for
this effect suitable for a compact model were later presented in [PKH01]. A bias
dependence of the Gummel number was derived, though without any discussion of
the model for the temperature effects. The derived model was included into the
compact model of the same authors ([TPK12]). It will be discussed briefly in Ch.
2.4.3.5, where it is compared to the model derived in this work, which is based on
the GICCR.

   This section first summarizes the relevant physical equations describing the transfer current and then discusses the influence of the Germanium profile on the transconductance and output conductance as well as on high injection and high current
effects.

2.4.1.1 Quasi-static transfer current equation

The formulation for the 1D transfer current in bipolar transistors is derived from
the drift-diffusion transport equation

$$- J_{\mathrm{n}} = q\overline{\mu_{\mathrm{nB}}}V_{\mathrm{T}}\frac{\mathrm{d}n}{\mathrm{d}x} + q\overline{\mu_{\mathrm{nB}}}nE_{\mathrm{nx}}, \qquad (2.21)$$

with the average electron mobility in the base $\overline{\mu_{\mathrm{nB}}}$ and the drift field of the electrons
E_{nx}. For simplification, a constant base doping is assumed leading for low injection
to

$$E_{\mathrm{nx}} = -\frac{\mathrm{d}(V_{\mathrm{p}} - V_{\mathrm{n}})}{\mathrm{d}x}, \qquad (2.22)$$

with the band potentials V_{n} and V_{p}. These band potentials contain all effects on the
edges of the conduction and valence band, including bandgap narrowing due to high
doping effects and bandgap changes caused by material compositions. Additionally,
changes of the effective mass are also included. Neglecting the latter and assuming

a linear change of the bandgap, the field can be written as

$$E_{nx} = -\frac{\Delta V_{G,max}}{\Delta x_{VG,max}}, \tag{2.23}$$

with the bandgap change in the base $\Delta V_{G,max}$ over the distant $\Delta x_{VG,max}$.

Solving the differential equation (2.21) leads to

$$n(x') = \left(n_e - \frac{J_n}{q\mu_{nB}E_{nx}}\right)\exp\left(-\frac{E_{nx}}{V_T}(x' - x_e)\right) + \frac{J_n}{q\mu_{nB}E_{nx}}. \tag{2.24}$$

Here, the origin of x' is defined at the BE junction x_{jE} such that $n_e = n(x_e)$ is the injected electron density at the beginning of the neutral base x_e.

In the following, the forward active operating region with sufficiently low $V_{B'C'}$ is assumed. Therefore, injection at the collector side of the base is neglected and the electrons travel with saturation velocity through the BC-SCR. Thus, the boundary condition at the beginning of the SCR reads.

$$n_c = n(x_c) = -\frac{J_n}{qv_{sn}}, \tag{2.25}$$

with the end of the neutral base x_c and the saturation velocity of electrons v_{sn}. With the width of the neutral base $w_B = x_c - x_e$, the drift-factor

$$\zeta = -\frac{E_{nx}w_B}{V_T}, \tag{2.26}$$

and

$$f_\zeta = \exp(\zeta), \tag{2.27}$$

the closed form solution of the transfer current is

$$I_T = -J_n A_{E0} = \frac{qA_{E0}n_e f_\zeta}{\frac{1}{v_{sn}} - \frac{f_\zeta - 1}{E_{nx}\mu_{nB}}}. \tag{2.28}$$

This formulation of the transfer current is introduced into (2.24), yielding a closed form solution of the electron density in the neutral base with

$$n(x') = n_e\left[(1 - r_\zeta)\exp\left(-\frac{E_{nx}}{V_T}(x' - x_e)\right) + r_\zeta\right], \tag{2.29}$$

and

$$r_\zeta = \frac{f_\zeta}{f_\zeta - 1 - \frac{\mu_{nB}E_{nx}}{v_{sn}}}. \tag{2.30}$$

The application of above equation is shown in Fig. 2.24 for different values of the drift factor ζ. A good approximation in most of the neutral base is achieved by this equation. The larger deviations especially for the large drift field are caused by the missing field dependence of the mobility. Visible in this picture and following also from (2.29) is a possible increase of the electron density compared to the linear drop given in the classical theory of the diffusion transistor.

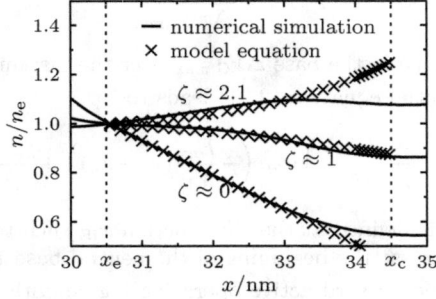

Figure 2.24: Electron density in the neutral base normalized to n_e from numerical device simulations and modeled by (2.29) for different values of ζ. Included are the boundaries of the neutral base.

While equation (2.28) is the general solution of the transport equation for the neutral base, a different solution is obtained for the case of a non-existing drift field in the base. The equation can be derived either by calculating the limit $E_{nx} \to 0$ in the above equation or by solving the differential equation (2.21) with $E_{nx} = 0$. In either case, the solution reads

$$I_T = \frac{q A_{E0} n_e}{\frac{1}{v_{sn}} + \frac{w_B}{\mu_{nB} V_T}}. \tag{2.31}$$

For high injection, i.e. $n \approx p \gg N_B$, the Webster effect [Web54] leads to a dependence of the hole density on the electron density and it follows for the electric field ([SC10])

$$E_{nx} = \frac{V_T}{n} \frac{dn}{dx} - \frac{d(V_p + V_n)}{dx}. \tag{2.32}$$

The term dn/dx leads together with (2.21) to a different solution of the transfer current. This is shown in Ch. 2.4.1.4.

2.4.1.2 Current gain and transconductance

Bias dependence
A strong bias dependence of the current gain was reported early for SiGe-HBTs ([CPS+90] and [CCP+93]). However, since the current gain also contains non-ideal effects of the base current, it is more useful to discuss a normalized transfer current[1] and the transconductance with the corresponding Reverse Early effect, since both are only related to the transfer current.

---

[1] The collector current which is equal to I_T in the selected operating region is shown in the plot.

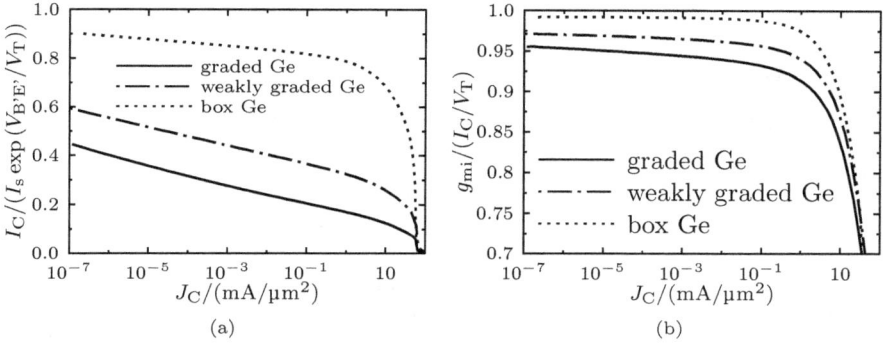

Figure 2.25: Bias dependence of (a) the normalized collector current $i_{C,norm}$ (cf. (2.37)) and (b) the normalized internal transconductance g_{mi} for different kinds of graded Germanium profiles in the base. Both have an ideal theoretical value of 1.

The Reverse Early effect describes the impact of the base width modulation on the transfer current at the emitter side of the base. For SiGe HBTs, a reduction of the transconductance is caused by a graded Germanium profile in the base. This effect was initially modeled in [PKH01]. A comparison for differently graded Germanium in the base of $i_{C,norm}$ and g_{mi} is given in Fig. 2.25. Showing there is a strong dependence of both on the shape of the Germanium content.

The physical explanation of the effect is based on the injection of minorities (electrons assumed here) into the base of the transistor. The intrinsic carrier density depends among others on the bandgap V_G and can thus be non-constant due to a changing bandgap. This change is caused in SiGe-HBTs by bandgap narrowing due to high doping effects and the Germanium mole fraction. Due to non-abrupt doping profiles and possible Germanium grading in the base, especially in the BE-SCR, V_G and thus n_i become a function of x'. A linear change of the bandgap is assumed, which is the most relevant case for practical application, leading to

$$n_i^2(x') = n_{i,jE}^2 \exp\left(\frac{x'}{a_{ni}}\right),\tag{2.33}$$

with $x' = x - x_{jE}$ and $n_{i,jE}$ as the intrinsic carrier density at $x = x_{jE}$. a_{ni} describes the slope of the Germanium content and is defined by

$$a_{ni} = -\frac{\Delta x_{Ge,max} V_T}{\Delta V_{G,max}}.\tag{2.34}$$

Note, a_{ni} is defined as positive value. For low injection, i.e. $n \ll p$, the injected electron density at the beginning of the neutral base is given by

$$n_e = \frac{n_i^2(x_e)}{N_B} \exp\left(\frac{V_{B'E'}}{V_T}\right).$$ (2.35)

Two cases have to be distinguished for practical transistor profiles. The first case assumes a strong grading of the Germanium only in the BE-SCR, while the mole fraction is constant with the maximum value in the neutral base. In the latter case, (2.31) is used to describe the transfer current. The other case assumes a weaker grading across the complete base, including the BE-SCR and the neutral base and uses (2.28). Although, if one assumes $\Delta x_{Ge,max} = x_{e0}$ for the first case, a portion the drift field will also be located in the neutral base for $V_{B'E'} > 0$. However, $w_B \gg x_{e0} - x_e$ is valid for all $V_{B'E'}$, allowing to neglect this small part. In both cases (2.33) is valid, since a grading of the Germanium content is located in the BE-SCR.

Inserting (2.35) with (2.33) into (2.28) and assuming a constant $\zeta = \zeta_0$, i.e. neglecting the classical Early effect caused by the base width modulation, leads to

$$I_s = q A_{E0} \frac{n_{i,jE}}{N_B} \exp\left(\frac{x_{e0}}{a_{ni}}\right) \frac{f_{\zeta 0}}{\frac{1}{v_{sn}} - \frac{f_{\zeta 0}-1}{E_{nx}\mu_{nB}}},$$ (2.36)

and

$$i_{C,norm} = \frac{I_T}{I_s \exp\left(\frac{V_{B'E'}}{V_T}\right)} = \exp\left(\frac{x_e - x_{e0}}{a_{ni}}\right).$$ (2.37)

The reduction of the normalized current shown in Fig. 2.25 is therefore directly correlated to the movement of the BE-SCR in the base, and also strongly depends on the gradient of the Germanium profile, i.e. a_{ni}.

The transconductance of a bipolar transistor is

$$g_{mi} = \left.\frac{dI_T}{dV_{B'E'}}\right|_{V_{C'E'}}.$$ (2.38)

For the graded Germanium in the base therefore follows

$$\frac{g_{mi}}{I_C/V_T} = 1 + \frac{V_T}{a_{ni}} \frac{dx_e}{dV_{B'E'}} + \frac{V_T}{a_{ni}} \left(1 - \frac{v_{sn}f_\zeta}{v_{sn}(f_\zeta - 1) - E_{nx}\overline{\mu_{nB}}}\right) \frac{dw_B}{dV_{B'E'}}.$$ (2.39)

Note, in the case of graded Germanium across the complete base, one can write

$$-\frac{E_x}{V_T} = \frac{1}{a_{ni}}.$$ (2.40)

In (2.39), the first term represents the ideal case while the last term describes the impact of the base width modulation. The second term explains the stronger g_{mi} reduction for graded Germanium transistors in a similar manner as in (2.37).

In summary, the increasing bandgap at x_e for increasing $V_{B'E'}$ and the thus decreasing intrinsic carrier density lead to a decreasing number of injected electrons normalized to $\exp(V_{B'E'}/V_T)$. Therefore, the gain of transistor performance by adding an aiding drift field to compensate e.g. retarding fields from slopes in the base doping is reduced by the lower transconductance. Also, the reduced transconductance may impact the nonlinearities of the transistor and therefore harmonic and intermodulation distortions.

Germanium profile estimation
Eq. (2.37) can also be used to estimate the Germanium profile in the BE-SCR. Since (2.34) can also be written in a differential form, it follows

$$i_{C,norm} = \exp\left(\frac{x_{e0}}{V_T}\frac{dV_G}{dx}(c-1)\right),\tag{2.41}$$

with

$$x_{e0} = \frac{\varepsilon N_E}{(N_E+N_B)C_{jEi0}},\tag{2.42}$$

and

$$c = \frac{x_e}{x_{e0}} = \frac{C_{jEi0}}{C_{jEi}}.\tag{2.43}$$

Rewriting (2.41) and integrating both sides allows to calculated V_G by

$$V_g(x_e) = V_g(x_{e0}) - \int_{x_{e0}}^{x_e(V_{B'E'})}\frac{\ln(i_{norm})V_T}{x_{e0}(c-1)}dx.\tag{2.44}$$

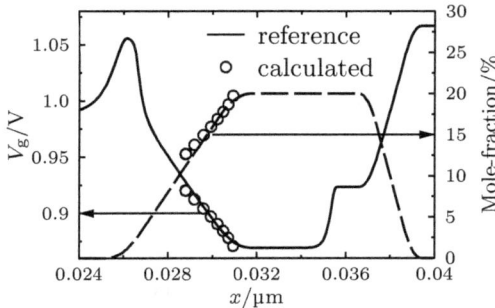

Figure 2.26: Calculated bandgap profile using (2.44). The start values of V_g and the mole fraction were taken from the known internal values.

This equation can be used to estimate the bandgap variation in the BE-SCR, starting from a known value $V_g(x_{e0})$. If this value is unknown, only the slope of the bangap, but not the absolute value can be calculated. However, since also the measured I_C at very low-bias can be quite noisy, the calculation of the slope can start from any value and is not limited to x_{e0}. Using the known dependence of the Germanium mole fraction on the bandgap, one can calculate the slope of the Germanium profile. If the absolute value of the mole fraction is not known at x_{e0}, the calculation only works when assuming a linear dependence on the bandgap. However, in the relative small region where this equation is applied, this function can always be linearized.

The application of (2.44) on 1D simulation results is shown in Fig. 2.26 for a strongly graded Germanium profile. Here, the value of the bandgap and the Germanium content for the reference location were taken from internal quantities of device simulations in order to allow a better comparison of the extracted slope. Also, the non-linear dependence of V_g on the Germanium content was linearized using 70 mV per 10% mole fraction. Using the simple assumptions, the extracted slope of the Germanium profile is close to the actual data from the simulation. However, it can be assumed that applying the equation on more weakly graded Germanium will reduce the achieved accuracy, since the impact of the yet neglected classical Early effect will be stronger.

Temperature dependence

In Fig. 2.27, the normalized collector current is shown for a large temperature sweep, showing the strong temperature dependence for the graded Germanium profile, while the values for the box profile stay almost constant.

Combining (2.37) with (2.34), one can write

$$\frac{i_{C,\text{norm}}(T)}{i_{C,\text{norm}}(T_0)} = \exp\left(\frac{x_e(T) - x_{e0}(T)}{a_{ni}(T)} - \frac{x_e(T_0) - x_{e0}(T_0)}{a_{ni}(T_0)}\right), \qquad (2.45)$$

which results in

$$\frac{i_{C,\text{norm}}(T)}{i_{C,\text{norm}}(T_0)} = \exp\left(\frac{x_{e0}(T_0)\left(\sqrt{1 - \frac{V_{B'E'}}{V_{DEi}(T_0)}} - 1\right)}{a_{ni}(T_0)}\right.$$
$$\left.\left(\frac{T_0}{T}\frac{C_{jEi0}(T_0)}{C_{jEi0}(T)}\frac{\sqrt{1 - \frac{V_{B'E'}}{V_{DEi}(T)}} - 1}{\sqrt{1 - \frac{V_{B'E'}}{V_{DEi}(T_0)}} - 1} - 1\right)\right). \qquad (2.46)$$

For a fixed $V_{B'E'}$, the normalized current increases with temperature, while its impact gets stronger with increasing bandgap change, i.e. decreasing a_{ni}. Similar results are obtained for the transconductance. This effect is further amplified for constant I_T, i.e. in curves as shown in Fig. 2.27, since $V_{B'E'}$ is decreasing with T and thus leads to an additional increase of $i_{C,\text{norm}}(T)$.

2.4.1.3 Output conductance

Bias dependence

In the previous section it was shown how the changing n_i at the end of the BE-SCR influences the transconductance of a transistor. Next, it is discussed how the value of ζ, i.e. the strength of the drift field, influences the output conductance of a transistor. Fig. 2.28 shows the output conductance normalized to the collector

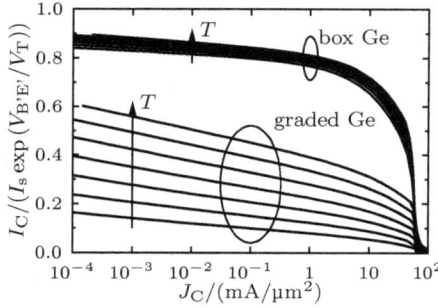

Figure 2.27: Normalized collector current for $T = [240\ldots420]\,\mathrm{K}$ for different Germanium profiles.

current as a function of the collector current density for the transistors with the electron density in the base as shown in Fig. 2.24. As one can see, the output conductance strongly depends on ζ and decreases for increasing ζ. Additionally, the normalized value depends on the operating point even for low injection, which is not the case for the pure diffusion transistor (i.e. $\zeta = 0$).

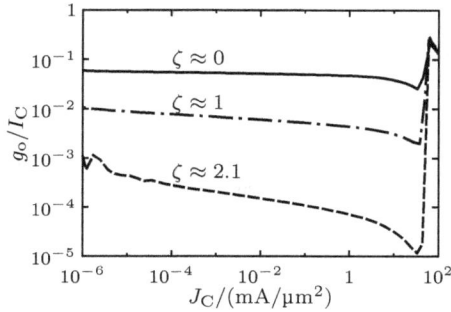

Figure 2.28: Output conductance normalized to the collector current as a function of collector current density for $V_{B'C'} = 0\,\mathrm{V}$ and different values of ζ. The shown bias range corresponds approximately to $V_{B'E'} = [0.5\ldots1.1]\,\mathrm{V}$.

Based on (2.28), the output conductance for low injection reads

$$g_{o} = \left.\frac{\mathrm{d}I_T}{\mathrm{d}V_{C'E'}}\right|_{V_{B'E'}} = \frac{I_T}{V_T}\left(E_{nx} - \frac{f_\zeta}{\frac{f_\zeta - 1}{E_{nx}} - \frac{\mu_{nB}}{v_{sn}}}\right)\frac{\mathrm{d}x_c}{\mathrm{d}V_{B'C'}}, \qquad (2.47)$$

with

$$\frac{dx_c}{dV_{B'C'}} = \frac{x_{jC} - x_{c0}}{2V_{DCi}\sqrt{1 - \frac{V_{B'C'}}{V_{DCi}}}}.$$

(2.48)

Note, in above equation $dx_c/dV_{B'C'}$ is positive. Assuming the electron mobility in the base is constant for different ζ and thus independent of the material composition, the drop of the normalized g_o is caused by f_ζ and E_{nx}.

Above equation predicts negative output conductances for large values of ζ. The border between positive and negative output conductances corresponds to the case, where the electron density shown in Fig. 2.24 is flat. Hence, if ζ is large enough to cause an increase of the electron density in the base rather than the decrease known from diffusion transistors, a negative Early effects follows from the theory. However, results of numerical device simulations show that the actual value of the output conductance will still be positive, though very small. The decrease of the electron density is compensated by a decrease of the mobility and an increase of the carrier velocity, since the latter is not equal but less than the saturation velocity. Therefore it is still a function of the electric field and compensating the negative g_o.

A large field corresponds to a large f_ζ. Therefore, for a very large drift field, i.e. $f_\zeta \to \infty$, g_o becomes zero, as shown in Fig. 2.28. In (2.47), the decrease of the normalized g_o with increasing I_T or $V_{B'E'}$, respectively, is caused by the increase of the neutral base with w_B, leading to an additional increase of f_ζ.

Temperature dependence

As shown in Fig. 2.29, in contrast to transistors with a constant Germanium profile in the base, where the output conductance is almost independent of the ambient temperature, it is a strong function of temperature for the graded profile. The value increases similar as already described for g_{mi} in 2.4.1.2.

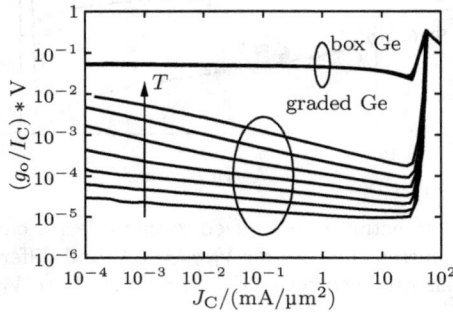

Figure 2.29: Normalized output conductance as a function of collector current for $T = [240\ldots420]$ K and different Germanium profiles.

Neglecting the temperature dependence of the parameters corresponding to the depletion capacitance in (2.48) and calculating the derivative of (2.47) normalized to I_T with respect to the temperature yields

$$\frac{\mathrm{d}(g_o/I_T)}{\mathrm{d}T} = \frac{1}{V_T}\frac{\mathrm{d}x_c}{\mathrm{d}V_{\mathrm{B'C'}}}\left[-\frac{1}{T}\left(E_{\mathrm{nx}} - \frac{f_\zeta}{\frac{f_\zeta-1}{E_{\mathrm{nx}}} - \frac{\overline{\mu_{\mathrm{nB}}}}{v_{\mathrm{sn}}}}\right) - \frac{1}{\frac{f_\zeta-1}{E_{\mathrm{nx}}} - \frac{\overline{\mu_{\mathrm{nB}}}}{v_{\mathrm{sn}}}}\frac{\mathrm{d}\zeta}{\mathrm{d}T}\right.$$

$$\left.+\frac{f_\zeta}{\left(\frac{f_\zeta-1}{E_{\mathrm{nx}}} - \frac{\overline{\mu_{\mathrm{nB}}}}{v_{\mathrm{sn}}}\right)^2}\frac{1}{E_{\mathrm{nx}}}\frac{\mathrm{d}\zeta}{\mathrm{d}T}\right] \qquad (2.49)$$

and is a function of the temperature due to $\mathrm{d}\zeta/\mathrm{d}T$.

2.4.1.4 Onset of high-injection effects

SiGe-HBTs partially show a strongly increased influence of high injection effects, i.e. the deviation of the transfer current its ideal slope for large currents. This is shown in Fig. 2.30 (a) especially for the "box"-HBT and the BJT. For the following discussion, only the forward active current I_{Tf} is discussed, which can be derived from (2.28) for low-bias when setting v_{sn} to ∞. The saturation current for low injection then reads with (2.35)

$$I_{\mathrm{Sfl}} = qA_E V_T \frac{\overline{\mu_{\mathrm{nB}}}}{w_B a_{\mathrm{Bfl}}}\frac{n_i^2(x_{e0})}{N_B^2}, \qquad (2.50)$$

with

$$a_{\mathrm{Bfl}} = \frac{f_\zeta - 1}{f_\zeta \zeta}. \qquad (2.51)$$

High injection effects start when the assumption of $n_e \ll N_B$ is not valid anymore. It is independent from high current effects. The general solution for the injected electron density is ([SC10])

$$n_e = \frac{N_B}{2}\left[\sqrt{1 + 4\frac{n_i^2(x_e)}{N_B^2(x_e)}\exp\left(\frac{V_{\mathrm{B'E'}}}{V_T}\right)} - 1\right], \qquad (2.52)$$

with the solution for very high injection

$$n_e = n_i(x_e)\exp\left(\frac{V_{\mathrm{B'E'}}}{2V_T}\right). \qquad (2.53)$$

From this, the saturation current for high injection reads ([SC10])

$$I_{\mathrm{Sfh}} = 2qA_E V_T \frac{\overline{\mu_{\mathrm{nBh}}}}{w_B a_{\mathrm{Bfh}}}n_i(x_{e0}), \qquad (2.54)$$

with

$$a_{\mathrm{Bfh}} = \frac{f_{\zeta h} - 1}{f_{\zeta h}\zeta_h}, \qquad (2.55)$$

and (for simplicity assuming $N_B = \text{const.}$)

$$\zeta_h = \frac{\zeta}{2}. \tag{2.56}$$

$f_{\zeta h}$ is defined similar to (2.27). $\overline{\mu_{nBh}}$ is the mean electron mobility in the base under high injection. Finally, the transfer current for very high injection reads

$$I_{Tf} = I_{Sfh} \exp\left(\frac{V_{B'E'}}{2V_T}\right). \tag{2.57}$$

Often the so-called Knee current I_{Kf} is used as a quantity describing the onset of high injection effects. When neglecting the Early effect, the transfer current is then written as

$$I_{Tf} = \frac{I_{Sfl}}{1 + \frac{I_{Tf}}{I_{Kf}}} \exp\left(\frac{V_{B'E'}}{V_T}\right). \tag{2.58}$$

For $I_{Tf} \gg I_{Kf}$ follows with (2.57) for I_{Kf}

$$I_{Kf} = \frac{I_{Sfh}^2}{I_{Sfl}}. \tag{2.59}$$

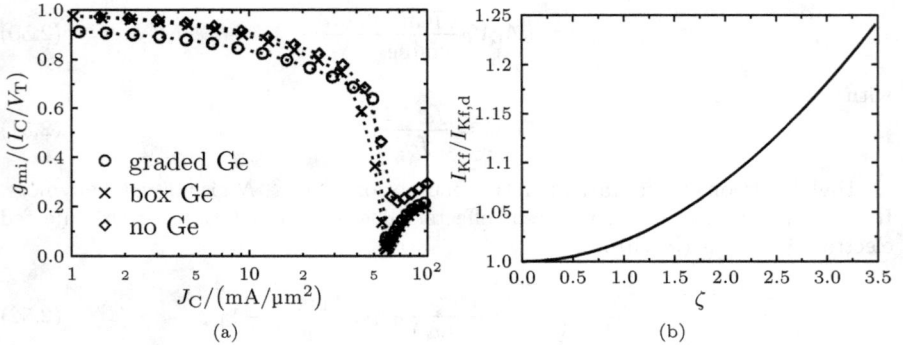

Figure 2.30: (a) Normalized transconductance as a function of collector current density for different Germanium profiles. (b) Knee current I_{Kf} normalized to its value for a diffusion transistor $I_{Kf,d}$ as a function of ζ.

Neglecting the difference between the $\overline{\mu_{nB}}$ from (2.50) and $\overline{\mu_{nBh}}$ from (2.54) I_{Kf} is expressed by

$$I_{Kf} = J_{TDl}N_B \frac{\zeta f_\zeta}{f_\zeta - \sqrt{4f_\zeta + 1}} \tag{2.60}$$

Here, J_{TDl} is the transfer current density normalized by $n_e(x_e)$. The application of this equation is given in Fig. 2.30, showing an increase of the knee current with increase electric field, though only a small dependence exists. Even at very high value of ζ, the increase with respect to the values for the diffusion transistor is only about 25%.

Although the increase of ζ also corresponds to an increase of $n_e(x_{e0})$ (cf. Ch. 2.4.1.2) and thus of the saturation current, the current corresponding to the Kirk effect, i.e. high current effects, remains almost the same. Therefore, the effects of high injection in the base are less pronounced with respect to high current effects for transistors with a high electric field.

The strong visible reduction of I_{Kf} for SiGe transistors must, thus, be caused by a different phenomenon. Additionally to the classical high injection effect in the base, also voltage drops across the neutral emitter and the BE-SCR affect the forward transfer current in a similar way as the high injection effects. Shown in Fig. 2.31 is the voltage drop in both regions for different Germanium profiles. In the neutral emitter it is identical at medium injection. However, especially for the transistor with the graded Germanium profile, the voltage drop in the BE-SCR is very high, about a factor of 4 with respect to the other. Thus, the reduced transit time in the base of the "grad"-BHT is achieved only by cost of a more non-ideal transfer current. The voltage drop in the neutral emitter differs for high currents for all transistors.

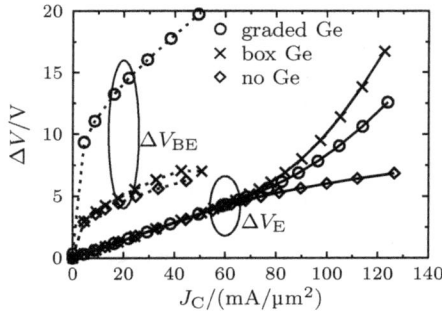

Figure 2.31: Voltage drop ΔV_E across the neutral emitter and ΔV_{BE} across the BE-SCR for different Germanium profiles. Note, the curve for ΔV_{BE} is only plotted up to the current density corresponding to the collapse of the SCR in order to simplify the plot. The plot is given for $V_{B'C'} = 0\,\mathrm{V}$.

The voltage drop across the neutral emitter depends on the electron density. From the Poisson equation, the latter can be expressed by the doping and the hole density in this region by

$$n = N_E + p_m, \qquad (2.61)$$

with the hole density in the neutral emitter p_m. Here, a model for p_m will not be derived, since it can be found in e.g. [SC10]. Also, this model is limited to low back injection in the emitter, since at high injection the same effects as for electrons in the base occur for the holes in the emitter. However, simple models for the current gain as given in e.g. [Cre06] show an increase with increasing Ge-content and thus a decrease of I_B at a given I_C. Since p_m is directly correlated to I_B, a lower number of mobile holes and electrons follows. For a given I_C, the ratio of minorities in the emitter for a Si and SiGe transistor is thus roughly given by

$$\frac{Q_{mE,SiGe}}{Q_{mE,Si}} = \frac{n_{i,Si}(x_{e0})}{n_{i,SiGe}(x_{e0})}. \tag{2.62}$$

The lower number of mobile electrons is compensated by a larger electric field at the same current density, leading to an increase of the voltage drop. As also given in [Cre06], the current gain is decreasing with increasing temperature. Therefore, the effect is less pronounced for higher temperatures, which also follows from (2.62).

The strong effect of the voltage drop in the emitter on the transconductance was also demonstrated for SiGe-HBTs in [Fri02] and for InP-DHBTs in [JR11].

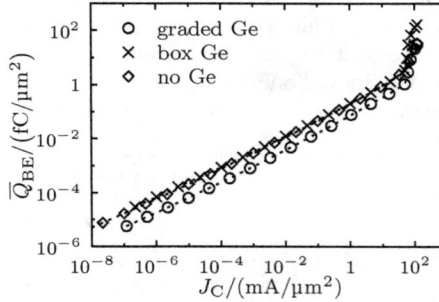

Figure 2.32: Minority charge in the BE-SCR for the profiles from Fig. 2.31.

The explanation for the voltage drop in the BE-SCR is also found in the different values of the charge in this region (cf. Fig. 2.32). This charge can be expressed as ([SC10])

$$Q_{BE} = \frac{Q_{BE0}}{(1 - \frac{V_{B'E'}}{V_{DEi}})^{(1/z_{Ei})}} \exp\left(\frac{V_{B'E'}}{2V_T}\right), \tag{2.63}$$

with

$$Q_{BE0} = \frac{2qA_E\overline{n_i}w_{BE0}V_T}{V_{DEi}}, \tag{2.64}$$

and an average value of the intrinsic carrier density in the BE-SCR $\overline{n_i}$. Since the term $\exp(V_{B'E'}/(2V_T))$ can be written as $\sqrt{I_T/I_s}$ when neglecting all non-idealities, $Q_{BE}(I_T)$ is almost equal for the box Germanium profile and the transistor without Germanium, since $\overline{n_i}$ in (2.64) the same as $n_i(x_{e0})$ in (2.50) and therefore canceling out. However, this holds only if the Ge-content of the HBT extends sufficiently far into the emitter.

This is not the case for the graded Germanium profile. Here, $\overline{n_i}$ is roughly equal to $n_{i,je}$ and therefore smaller than $n_i(x_{e0})$. Similar as in (2.62), a ratio of the charges at constant I_T can be derived. In this case, the ratio is meaningful between graded and non-graded Germanium content and is given roughly by

$$\frac{Q_{\mathrm{BE,grad}}}{Q_{\mathrm{BE,box}}} = \frac{n_{i,\mathrm{je}}}{n_i(x_{\mathrm{e0}})}. \tag{2.65}$$

2.4.1.5 Operation at high current densities

At high current densities, the reduced drift field in the BC-SCR cannot carry electrons with saturation velocity anymore. A diffusion current in the collector is required to support the current density at the end of the base. This leads to an increase of the electron density and due to the neutral injection zone also to an increase of the hole density. However, due to the barrier holes are accumulating at the heterojunction in the collector for SiGe transistors. Thus, also the electron density is decreasing, reducing the current flow in the injection zone.

The effect is visualized in Fig. 2.33, showing two transistors with the same BE-junction and thus the same injection in the base. However, one has a heterojunction in the collector ("box" transistor) while the other does not. Note that obviously the "Si" transistor is not a pure BJT (Ge content in the complete collector) but the collector junction follows the theory of the homojunction. It can clearly be seen, that for the same injection in the base the carrier density in the collector is much lower for the heterojunction. This has direct impact on the transfer current, where the high current effects are much more pronounced for the transistor with the heterojunction.

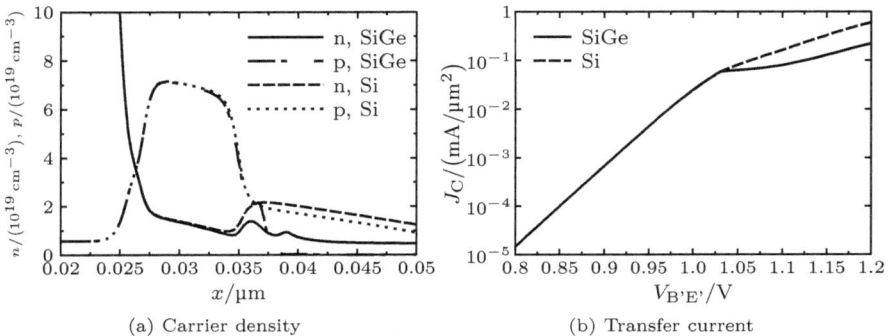

Figure 2.33: (a) Electron and hole density at the beginning of high current effects. (b) Transfer current. Both plots are for transistors with the same BE-junction but with a heterojunction in the collector ("SiGe") and without ("Si").

2.4.1.6 Summary

In the previous section it was shown that the introduction of Germanium, either with graded or with a box profile, adds several aspects to the most important current characteristics of a bipolar transistor in forward active operation. Corresponding effects comprise the introduction of the bias and temperature dependence of those characteristics, which are independent of the latter for pure Silicon based technologies. The effects are all correlated to the gradient of the Germanium fraction and the differences in the bandgap of the Silicon and SiGe region, with reduced influence at higher temperature.

It is therefore necessary to extend existing compact model formulations by these effects, in order to enable accurate modeling and thus circuit simulations. The GICCR is ideally suited for this purpose, since the partitioning into physics-based components enables improvements correlated to the physical origin of shown effects.

2.4.2 Modeling with the GICCR

2.4.2.1 Introduction to the GICCR

Already in early publications (e.g. [Gum64] and [GDLMD67]), a closed form analytical expression for the transfer current based on the hole density in the base was derived. In [SFR93], the integration boundaries for the holes were moved to the terminals of the complete 1D transistor, i.e. the emitter and collector contact rather than the neutral base. Also a partitioning of the integrand based on the so called weight factors was introduced.

Based on rewriting the drift-diffusion transport equation, the closed form solution of the transfer current in a 1D transistor reads

$$I_T = \frac{c_0}{\int_0^{l_{EC}} h_g(x) h_J(x) h_v(x) p(x) dx} \left[\exp\left(\frac{V_{B'E'}}{V_T}\right) - \exp\left(\frac{V_{B'C'}}{V_T}\right) \right], \qquad (2.66)$$

with the GICCR constant

$$c_0 = q V_T \overline{\mu_{nr} n_{ir}^2}, \qquad (2.67)$$

the depth of the transistor l_{EC} and weight functions

$$h_g(x) = \frac{\overline{\mu_{nr} n_{ir}^2}}{\mu_x(x) n_i^2(x)} \quad , \quad h_J(x) = -\frac{A_E J_n(x)}{I_T} \quad \text{and}$$

$$h_v(x) = \exp\left(\frac{V_{B'E'} - \varphi_p(x)}{V_T}\right). \qquad (2.68)$$

Since in all relevant operating ranges, $h_J \approx 1$ and $h_v \approx 1$, a simplified weight function

$$h(x) = \frac{\overline{\mu_{nr} n_{ir}^2}}{\mu_x(x) n_i^2(x)} \qquad (2.69)$$

is defined. The reference location to obtain the value of $\overline{\mu_{nr} n_{ir}^2}$ can be chosen arbitrarily. A comprehensive derivation of these equations and a comparison of results from different integration boundaries is given in e.g. [SC10].

In order to make (2.66) applicable for compact models, a charge partitioning scheme is applied, dividing the weighted hole charge into depletion and minority charges. The final equation after partitioning of the charge and calculating the respective weight factors therefore reads

$$Q_{p,T} = h_0 Q_{p0} + h_{BE} Q_{jEi} + h_{BC} Q_{jCi} + Q_{m,T}. \tag{2.70}$$

Here,

- Q_{p0} is the zero-bias hole charge in the transistor and h_0 the corresponding weight factor.

- $Q_{j(E,C)i}$ are the depletion charges corresponding to the BE- and BC-SCR, with the weight factors $h_{B(E,C)}$.

- $Q_{m,T}$ is the weighted minority charge in the transistor. This charge is further partitioned according to neutral regions and space charge region, yielding the charges $Q_{m(E,BE,B,BC,C)}$ with the weight factors $h_{m(E,BE,B,BC,C)}$. After onset of the high current effects, the collector charge can be further divided into the neutral collector and the injection zone.

The general equation to derive a weight factor for a specific charge is

$$h_R = \frac{\int_R h(x) p(x) \mathrm{d}x}{\int_R p(x) \mathrm{d}x} = q \frac{\int_R h(x) p(x) \mathrm{d}x}{Q_R}, \tag{2.71}$$

where R is an arbitrary location in the transistor with the corresponding hole charge Q_R. Often, above equation is further simplified to

$$h_R = \frac{\overline{\mu_{nr} n_{ir}^2}}{\overline{\mu_{nR} n_{iR}^2}}, \tag{2.72}$$

defining an effective value of $\overline{\mu_{nR} n_{iR}^2}$ for this transistor region.

In recent model approaches using the GICCR (e.g. [Sch05], [SC10]), all weight factors are assumed to be bias and temperature independent. However, as shown in the next chapters, modeling the effects caused by the introduction of Germanium into the Silicon base requires accurate models for both, the bias and temperature dependence.

2.4.2.2 Limits of the GICCR

As shown in e.g. [SFR93] and [Fri02], the voltage drop in the neutral emitter is correctly included in the GICCR when moving the integration boundary to the emitter contact. Further, in App. A.2.1 the general equivalence between GICCR and voltage drops, not only limited to the emitter region, is shown.

The influence of the voltage drop in the neutral emitter is included in the model for the transfer current, though *not* in that of the base current for standard compact models like HICUM/L2. The latter is always in a form similar to

$$I_{\mathrm{B}} = I_{\mathrm{BEs}} \exp\left(\frac{V_{\mathrm{B'E'}}}{m_{\mathrm{BE}} V_{\mathrm{T}}}\right), \qquad (2.73)$$

with the non-ideality factor m_{BE} that, however, only takes effects at low injection into account.

Generally the mono-emitter is highly doped[1]. Therefore, the resistance will not dominate over the poly-emitter and the contact region. However, due to increasing optimization of process flows, this ratio can shift towards the mono-emitter in the future.

2.4.2.3 Weight factor of the zero-bias charge

Using (2.72), the weight factor for Q_{p0} is expressed by

$$h_0 = \frac{\overline{\mu_{\mathrm{nr}} n_{\mathrm{ir}}^2}}{\overline{\mu_{\mathrm{nB}} n_{\mathrm{iB}}^2}}, \qquad (2.74)$$

with the constant value $\overline{\mu_{\mathrm{nr}} n_{\mathrm{ir}}^2}$ and the average value $\overline{\mu_{\mathrm{nB}} n_{\mathrm{iB}}^2}$ in the neutral base. In contrast to the numerator that comes from the constant (2.67), the denominator is bias dependent due to the field dependence of the electron mobility in the base. This bias dependence leads to a non-constant value of h_0, as shown in Fig. 2.34 for transistors with graded Germanium in the base and a box Germanium profile.

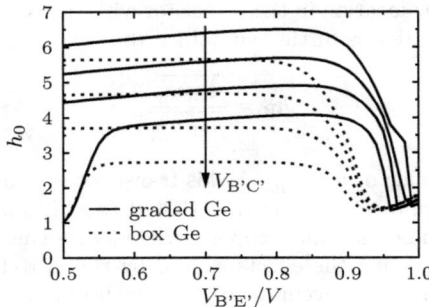

Figure 2.34: Bias dependence of the weight factor for the zero-bias hole charge as a function of $V_{\mathrm{B'E'}}$ for $V_{B'C'} = [-1.0, -0.5, 0, 0.5]\,\mathrm{V}$.

For the graded Germanium profile a dependence on both terminal voltages was found. In the low injection region, h_0 is increasing with $V_{\mathrm{B'E'}}$ but decreasing with

---

[1] For LEC profiles, the lightly doped region is generally completely depleted.

$V_{B'C'}$. While the dependence on $V_{B'E'}$ disappears for the box Germanium profile, the dependence on $V_{B'C'}$ still exists.

The bias dependence of h_0 with respect to $V_{B'C'}$ is mainly caused by the changing electric field and, thus, the also changing electron mobility. The mobility depends on the electric field by

$$\mu_n = \frac{\mu_{n0}}{\left[1 + \left(\frac{|E|}{E_{\lim}}\right)^\beta\right]^{1/\beta}}. \tag{2.75}$$

For large electric fields, i.e. $|E| \gg E_{\lim}$, the equation simplifies independent of β to

$$\frac{\mu_{n0}}{\mu_n} = \frac{|E|}{E_{\lim}}, \tag{2.76}$$

leading to a linear increase of the inverse value of the mobility with increasing field.

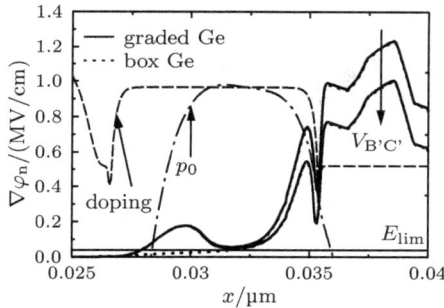

Figure 2.35: Electric field in the base and both SCRs for different Germanium profiles. For reference, the doping profile and the hole concentration at zero volt are included as well as $E_{\lim}$. Curves are shown for $V_{B'E'} = 0.8\,\text{V}$ and $V_{B'C'} = [0, -0.5]\,\text{V}$.

The spatial dependence of the electric field in the neutral base and the BC-SCR for different $V_{B'C'}$ is shown in Fig. 2.35. Obviously, the electric field in the BC-SCR increases with larger reverse bias. For base widths of approx. 10 nm, the value of the electric field is always much larger compared to $E_{\lim}$. This allows to insert (2.76) into (2.69), giving

$$h(x) = \frac{\overline{\mu_{nr} n_{ir}^2}}{\mu_{n0} n_{iB}^2(x)} \frac{|E(x)|}{E_{\lim}}. \tag{2.77}$$

For the derivation of a model equation it is useful to define an alternative x' as $x' = x_{jC} - x$, i.e. a reversion and shift to the BC junction. For simplification, a graded Germanium profile across the complete base is assumed, i.e. one can use (2.33) with (2.34) and define $\Delta x_{Ge,max} = x_{jC} - x_{jE}$. However, due to the shifted x-axis, $n_{i,jC} = n_i(x' = 0)$ is used.

Furthermore, a box like profile for p_0 is assumed between x'_{e0} and x'_{c0}. The latter assumption makes the following derivation only valid for reverse bias, but is sufficient for qualitative analysis. Using

$$E_{jC}(V_{B'C'}) = \sqrt{\frac{2q}{\varepsilon} \frac{N_B N_{Ci}}{N_B + N_{Ci}} (V_{DCi} - V_{B'C'})}, \qquad (2.78)$$

and

$$x'_c(V_{B'C'}) = \sqrt{\frac{2\varepsilon}{q} \frac{N_{Ci}}{N_B (N_{Ci} + N_B)} (V_{DCi} - V_{B'C'})}, \qquad (2.79)$$

gives the spatial dependence of the electric field in the BC-SCR with

$$E(x') = \frac{E_{jC}}{x'_c} (x'_c - x') = \frac{qN_B}{\varepsilon} (x'_c - x'). \qquad (2.80)$$

In the neutral base, a mobility of $\mu_{nB} = \mu_{n0}$ is assumed, which corresponds according to (2.76) to $E = E_{lim}$. Therefore, h_0 is calculated by

$$h_0(V_{BC}) = c_{h0} \left[\int_{x'_{c0}}^{x'_c} \exp\left(\frac{x'}{a_{ni}}\right) \frac{E_{lim} + E(x')}{E_{lim}} dx' + \int_{x'_c}^{x'_{e0}} \exp\left(\frac{x'}{a_{ni}}\right) dx' \right], \quad (2.81)$$

with

$$c_{h0} = A_E \frac{\overline{\mu_{nr} n_{ir}^2}}{n_{i,jC}^2 \mu_{n0}} \frac{qN_B}{Q_{p0}}. \qquad (2.82)$$

Note, E_{lim} in the numerator of the first integral in (2.81) is added to avoid discontinuities in the modeled electric field between the first and the second integral. Inserting (2.80) into (2.81) results in

$$h_0(V_{B'C'}) = c_{h0} a_{ni} \exp\left(\frac{x'_{c0}}{a_{ni}}\right) \left[\frac{qN_B a_{ni}}{\varepsilon E_{lim}} \left(\left(\frac{x'_{c0} - x'_c}{a_{ni}} - 1\right) + \exp\left(\frac{x'_c - x'_{c0}}{a_{ni}}\right) \right) + \left(\exp\left(\frac{x'_{e0} - x'_{c0}}{a_{ni}}\right) - 1 \right) \right]. \qquad (2.83)$$

Note, this derivation is only a rough model for h_0 and valid only for reverse bias due to the made assumptions. However, as it shows the correct dependence on $V_{B'C'}$ in the assumed bias region, it is sufficient here. A more accurate equation can be derived when using a more elaborate model for p_0 and the inverse mobility.

For the same reasons, it is not possible here to present a simple model for the dependence on $V_{B'E'}$. In the relevant operating range, the BE junction is forward biased, leading to a shift of the electric field outside of the neutral base from equilibrium. This also explains the almost not existing $V_{B'E'}$ dependence of the box profile as shown in Fig. 2.34. The small bias dependence of the graded profile is caused by the real spatial dependence of p_0, which is not box-like as assumed here. It was explained in 2.4.1.2 that the graded Germanium causes a decrease of n_i towards

x_{jE}. Thus, with increasing $V_{B'E'}$, the field moves into regions of low n_i, causing an increase of h_0.

Normalization

In order to reduce the number of parameters when applying the GICCR to compact models, the transfer current equation is normalized to the weight factor of the zero-bias charge h_0.

In general, h_0 can be expressed by

$$h_0 = h_{00} + \Delta h_{0,e} + \Delta h_{0,c}, \tag{2.84}$$

with the zero-bias value h_{00} and $\Delta h_{0,e}$ and $\Delta h_{0,c}$ as the change with respect to $V_{B'E'}$ and $V_{B'C'}$. Since in the final model equation the value in the numerator is still a constant, the normalization is performed by h_{00} rather than h_0

$$c_{10} = \frac{c_0}{h_{00}}, \tag{2.85}$$

leading in the low-bias region to

$$I_T = \frac{c_{10}}{Q_{p0} + \frac{\Delta h_{0,e}}{h_{00}} Q_{p0} + h'_{jEi} Q_{jEi} + \frac{\Delta h_{0,c}}{h_{00}} Q_{p0} + h'_{jCi} Q_{jCi}}$$
$$\left[\exp\left(\frac{V_{B'E'}}{V_T} \right) - \exp\left(\frac{V_{B'C'}}{V_T} \right) \right], \tag{2.86}$$

with the intermediate normalized values

$$h'_{jEi} = \frac{h_{BE}}{h_{00}} \text{ and } h'_{jCi} = \frac{h_{BC}}{h_{00}}. \tag{2.87}$$

Since the influence of the bias dependent changes of h_0 on $Q_{p,T}$ cannot be distinguished from the influence of the weighted depletion charges, the original form of the equation is maintained. This reads

$$I_T = \frac{c_{10}}{Q_{p0} + h_{jEi} Q_{jEi} + h_{jCi} Q_{jCi}} \left[\exp\left(\frac{V_{B'E'}}{V_T} \right) - \exp\left(\frac{V_{B'C'}}{V_T} \right) \right], \tag{2.88}$$

with the "terminal"[1] low-bias weight factors

$$h_{jEi} = h'_{jEi} + \frac{\Delta h_{0,e}}{h_{00}} \frac{Q_{p0}}{Q_{jEi}} \text{ and } h_{jCi} = h'_{jCi} + \frac{\Delta h_{0,c}}{h_{00}} \frac{Q_{p0}}{Q_{jCi}}. \tag{2.89}$$

The different values of the weight factors are summarized in Fig. 2.36. For h_{jEi}, the bias dependence of h_0 only results in a small shift of the value. In contrast, the value of h_{jCi} differs by orders of magnitude from that of h_{BC}[2]. The reference given

---

[1] They are called "terminal" weight factors since these are the values seen from extraction by using terminal quantities.

[2] This is also true for BJTs and not a phenomenon related to the heterojunctions.

here is based on calculating the weighted charge by

$$h_{jEi} = \left. \frac{Q_{p,T} - Q_{p0}}{Q_{jEi}} \right|_{V_{B'C'}=0\,V}, \tag{2.90}$$

and

$$h_{jCi} = \frac{Q_{p,T} - Q_{p,T}(V_{B'C'} = 0\,V)}{Q_{jCi}}, \tag{2.91}$$

with

$$Q_{p,T} = \frac{c_{10}}{I_C} \left[\exp\left(\frac{V_{B'E'}}{V_T}\right) - \exp\left(\frac{V_{B'C'}}{V_T}\right) \right], \tag{2.92}$$

and correspond to the values which will be extracted by dedicated extraction methods (cf. Ch. 3.3.1).

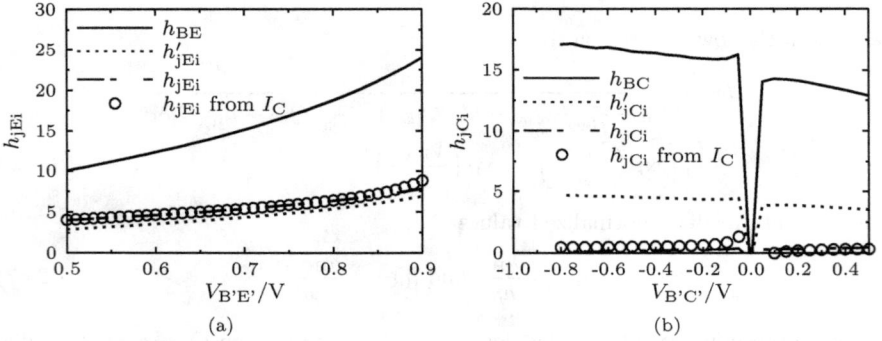

Figure 2.36: Bias dependence of (a) h_{jEi} for the graded Ge profile and (b) h_{jCi} for the box Ge profile. Values calculated from internal data (h_{BE} and h_{BC}), the normalized values defined in (2.87) and the final effective values taking the bias dependence of h_0 into account as given in (2.89) are included. For reference, the extracted values based on I_C are also given.

2.4.2.4 BE depletion charge related weight factor

As shown in Fig. 2.37, the effect of the g_{mi} degradation as explained in Ch. 2.4.1.2 is reflected in the bias dependence of the weight factor h_{BE}. The strong g_{mi} reduction from the profile with the strongly graded Germanium in the BE-SCR is directly correlated to a strong bias dependence, while it is the other way around for the box profile.

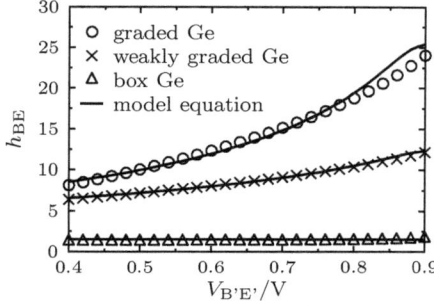

Figure 2.37: Bias dependence of the weight factor h_{BE} as a function of $V_{B'E'}$ for different gradings of the Germanium profile in the BE-SCR. Shown here is the comparison between results from REGAP and the application of (2.97) at $V_{B'C'} = 0\,\mathrm{V}$.

The weight factor h_{BE} is calculated from

$$h_{BE} = \frac{\int_{x_e}^{x_{e0}} h(x)p(x)\mathrm{d}x}{\int_{x_e}^{x_{e0}} p(x)\mathrm{d}x}. \tag{2.93}$$

For deriving a suitable compact model equation a constant $p(x) = N_B$ is assumed and

$$h(x) = \frac{\overline{\mu_{nr}n_{ir}^2}}{\overline{\mu_{nBE}}n_i^2(x)} \tag{2.94}$$

where (2.33) is employed. Moreover, the mobility in BE-SCR $\overline{\mu_{nBE}}$ is also assumed to be constant. Combining (2.93) with (2.94) allows to write

$$h_{BE} = c_{hBE}\frac{\int_{x_e}^{x_{e0}} \exp\left(-\frac{x'}{a_{ni}}\right)}{x_{e0} - x_e}, \tag{2.95}$$

with the constant value

$$c_{hBE} = \frac{\overline{\mu_{nr}n_{ir}^2}}{\overline{\mu_{nBE}}n_{i,jE}^2}. \tag{2.96}$$

The solution of (2.95) reads

$$h_{BE} = h_{BE0}\frac{\exp\left(u_{hBE}\right) - 1}{u_{hBE}}, \tag{2.97}$$

with

$$u_{hBE} = a_{hjEi}\left(\frac{x_e}{x_{e0}} - 1\right) = a_{hjEi}\left(1 - \sqrt{1 - \frac{V_{B'E'}}{V_{DEi}}}\right), \tag{2.98}$$

and the value extrapolated to $V_{B'E'} = 0\,V$

$$h_{BE0} = c_{hBE} \exp\left(-a_{hjEi}\right),\qquad(2.99)$$

which is a model parameter. The additional model parameter describing the strength of the bias dependent increase of h_{BE} is

$$a_{hjEi} = \frac{x_{e0}}{a_{ni}},\qquad(2.100)$$

and therefore proportional to the bandgap change. The application of this model is also shown in Fig. 2.37, providing a good agreement for all selected Germanium profiles. Note for the box profile that the weight factor is kept constant.

Both parameters are temperature dependent. For $h_{BE0}(T)$ a temperature independent mobility is assumed. Hence, h_{BE0} depends only on the ratio $\overline{n_{ir}^2}/n_{i,jE}^2$ and a_{hjEi}, yielding the equation for the temperature dependence

$$h_{BE0}(T) = h_{BE0}(T_0) \exp\left(\frac{\Delta V_{gBE}}{V_T}\left(\frac{T}{T_0} - 1\right)\right) \exp\left(a_{hjEi}(T_0) - a_{hjEi}(T)\right).\qquad(2.101)$$

Based on (2.100) and (2.34), the temperature dependence of a_{hjEi} is

$$a_{hjEi}(T) = a_{hjEi}(T_0)\frac{T_0}{T}\frac{C_{jEi0}(T_0)}{C_{jEi0}(T)}.\qquad(2.102)$$

The application of both equations is shown in Fig. 2.38 showing a good agreement in the given temperature range.

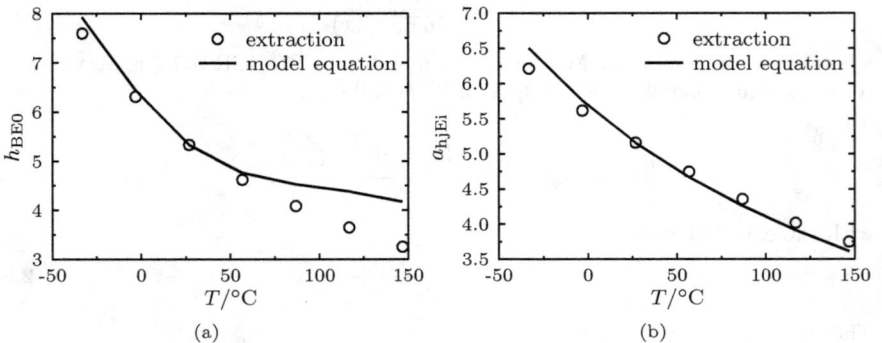

Figure 2.38: Models for the temperature dependences of (a) h_{BE0} according to (2.101) and (b) a_{hjEi} according to (2.102) compared to values extracted by applying a least square optimization of (2.97) on REGAP data.

Based on the results shown in sec. 2.4.2.3, h_{jEi} is derived from h_{BE} by scaling h_{BE0} with h_{00} and adding a small shift due to $h_0(V_{B'E'})$. The latter is assumed to be rather small, allowing to still use (2.97) for h_{jEi} with

$$h_{jEi}(V_{B'E'}) = h_{jEi0} \frac{\exp(u_{BE}) - 1}{u_{BE}}, \tag{2.103}$$

and the model parameter h_{jEi0} and the same u_{BE} as in (2.98). Due to the shift, the value of the other model parameter a_{hjEi} is slightly altered compare to (2.100), but still close to the physical value. The results of this model are later shown in 2.4.3.1, where the complete implementation of the model is summarized.

2.4.2.5 BC depletion charge related weight factor

The model for the weight factor h_{BC} is derived by using the same assumptions for the electric field as in Ch. 2.4.2.3. The equation reads

$$h_{BC} = c_{hBC} \frac{\int_{x'_{c0}}^{x'_c} \exp\left(\frac{x'}{a_{ni}}\right) \frac{E_{lim} + E(x')}{E_{lim}} \, dx'}{\int_{x'_{c0}}^{x'_c} dx'}, \tag{2.104}$$

with x' as defined in Ch. 2.4.2.3 and

$$c_{hBC} = \frac{\overline{\mu_{nr} n_{ir}^2}}{n_{i,jC}^2 \mu_{n0}}. \tag{2.105}$$

The solution is

$$h_{BC} = h_{BC0} \frac{c_E \left(\exp(u_{hBC}) - (u_{hBC} + 1)\right) + \exp(u_{hBC}) - 1}{u_{hBC}}, \tag{2.106}$$

with a prefactor for the field dependent term

$$c_E = \frac{q N_B a_{ni}}{E_{lim} \varepsilon}, \tag{2.107}$$

and the zero-bias value

$$h_{BC0} = c_{hBC} \exp\left(a_{hjCi}\right), \tag{2.108}$$

with

$$a_{hjCi} = \frac{x'_{c0}}{a_{ni}}. \tag{2.109}$$

The bias dependence of the SCR is described by

$$u_{hBC} = a_{hjCi} \left(\frac{x'_c}{x'_{c0}} - 1\right) = a_{hjCi} \left(1 - \sqrt{1 - \frac{V_{B'C'}}{V_{DCi}}}\right). \tag{2.110}$$

Except for the term caused by the electric field the solution is similar to the model for h_{BE}, resulting in similar equations for the temperature dependence, with

$$h_{BC0}(T) = h_{BC0}(T_0) \exp\left(\frac{\Delta V_{gBC}}{V_T}\left(\frac{T}{T_0} - 1\right)\right) \exp\left(a_{hjCi}(T_0) - a_{hjCi}(T)\right). \quad (2.111)$$

and

$$a_{hjCi}(T) = a_{hjCi}(T_0)\frac{T_0}{T}\frac{C_{jCi0}(T_0)}{C_{jCi0}(T)}. \quad (2.112)$$

The bandgap difference inserted in (2.111) is now positive, leading to an increase of h_{BC0} with temperature in contrast to the decrease of h_{BE0}.

To calculate the value of h_{jCi} according to the normalization explained in sec. 2.4.2.3, (2.83) and (2.106) are inserted into (2.89). Furthermore, (2.83) is rewritten with the variables from this section into

$$h_0 = h_{00}\left[c_E\frac{\exp(u_{hBC}) - (u_{hBC} + 1)}{\exp(u_{h0}) - 1} + 1\right], \quad (2.113)$$

with

$$h_{00} = \frac{h_{BC0}}{u_{h0}}\left[\exp(u_{h0}) - 1\right], \quad (2.114)$$

and

$$u_{h0} = a_{hjCi}\left(\frac{x'_{e0}}{x'_{c0}} - 1\right). \quad (2.115)$$

The final result for h_{jCi} is

$$h_{jCi} = h_{jCi0}\frac{\exp(u_{hBC}) - 1}{u_{hBC}}, \quad (2.116)$$

with

$$h_{jCi0} = \frac{u_{h0}}{\exp(u_{h0}) - 1}. \quad (2.117)$$

The terms caused by the electric field cancel out and a model similar to the model of h_{BE} is obtained. With increasing gradient of the Germanium profile a_{ni} and thus u_{h0} are increasing, leading to a decrease of h_{jCi}, which corresponds to the results from 2.4.1.3, i.e. a decreasing transconductance with increasing electric field.

Since the model equation of h_{jCi0} is the same as for h_{jEi0}, the same temperature model (though with a possibly different value of the bandgap difference) is applied. Results are shown in Fig. 2.39. Note, h_{jCi0} is increasing with temperature, because for a grading across the complete base, the bandgap is lower at x_{jC} than in the neutral base. For h_{jEi0}, the difference has the opposite sign. However, in case the grading is only in the BE-SCR or the Germanium follows a box profile, h_{jCi0} is almost constant with T.

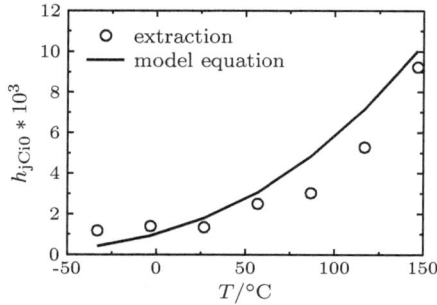

Figure 2.39: Model for the temperature dependence of h_{jCi0} for a transistor with graded Germanium in the base.

2.4.2.6 Weight factor for the mobile charge

Larger voltage drops in the neutral emitter and the BE-SCR exist for transistors with Germanium content in the base. In App. A.2.1 the correlation between voltage drops and weighted charges is discussed. From (A.28) it can be derived that a larger voltage drop correlates to a larger weighted charge. However, the larger voltage drop itself is caused by a lower charge, which finally results in a strong increase of the weight factor. From this, the square of n_i in the weight function (2.69) can be visualized, because the weight factor does not only compensate the lower charge but must also relate to the larger voltage drop.

In the low and medium current region, the transit time and corresponding minority charges are associated with the neutral emitter, the BE-SCR, the neutral base and the charge in the BC-SCR. The effects of the first three components were already discussed Ch. 2.4.1.4, while the effect of the latter is discussed here.

Base-collector SCR
The electron density in the BC-SCR is given by (2.25). Increasing the current reduces the space charge on the collector side by increasing the electron density, which can only be compensated by holes in the base. Therefore, a current dependence of the BC depletion charge exists. In compact models without a complete model for the electric field in the BC-SCR (cf. 2.5), this current dependence, which in turn results in a transit time, is generally modeled also by the minority transit time.

A charge ΔQ_{pBC} is defined as

$$\Delta Q_{pBC} = Q_{pBC} - Q_{jCi}, \tag{2.118}$$

with Q_{pBC} as the complete depletion charge in the BC-SCR, and therefore still a compensation charge with respect to Q_{p0}. Q_{jCi} corresponds to the charge from the classical current independent *theory* of the depletion charge.

The corresponding weight factor $h_{\Delta pBC}$ follows the theory of h_{BC}, since ΔQ_{pBC} is located in the same location as Q_{jCi}. With increasing current the space charge

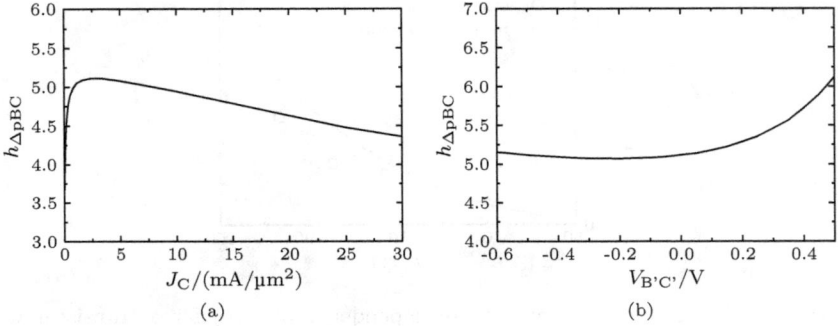

Figure 2.40: Weight factor of the compensating hole charge of the BC-SCR (a) versus J_C at $V_{B'C'} = 0\,V$ and (b) versus $V_{B'C'}$ at $J_C = 10\,\text{mA}/\mu\text{m}^2$.

is always decreasing in this region. Hence, ΔQ_{pBC} always has a positive sign. As a consequence, (2.106) cannot be used for calculating $h_{\Delta pBC}$ since the conditions are outside of the validity range. However, as shown in Fig. 2.40, values are fairly constant with respect to I_C and $V_{B'C'}$, which is in alignment with the theory that the corresponding holes are located in the neutral base and hence outside of the large electric field in the SCR.

Base-emitter SCR

The model for h_{mBE} follows directly from (2.65) as

$$h_{mBE} = \frac{\overline{\mu_{nr} n_{ir}^2}}{\mu_{nBE0} n_{i,je}^2}. \tag{2.119}$$

It thus corresponds to the bandgap change between the neutral base and the BE-SCR. Values for h_{mBE} obtained from numerical simulations are given in Fig. 2.41, showing similar values for the box profile and the transistor without Germanium in the base. This correlates to $n_{ir} \approx n_{i,je}$. However, for the transistor with the graded Germanium profile, $n_{i,je}$ is much smaller than n_{ir}, leading to a strongly increased h_{mBE}. At medium current densities, the value of h_{mBE} is roughly constant and thus does not require a dedicated model for the bias dependence.

Neutral emitter

In Fig. 2.31, the voltage drop across the neutral emitter before the onset of high injection into the base was shown to be the same, independent of the Germanium profile. Though, as further shown in the corresponding section, the hole charge strongly differs. Evaluating the weight function (2.69) in the neutral emitter, one obtains

$$h_{mE} = \frac{\overline{\mu_{nr} n_{ir}^2}}{\mu_{nE0} n_{iE}^2}, \tag{2.120}$$

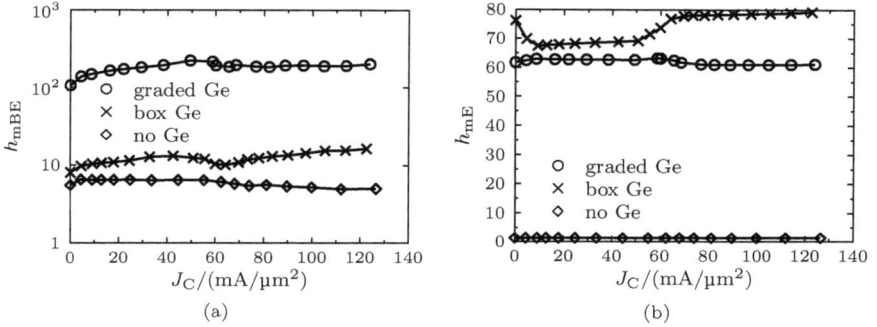

Figure 2.41: Weight factor for (a) the neutral charge in the BE-SCR and (b) the minority charge in the emitter as a function of J_C at $V_{B'C'} = 0\,\text{V}$ for different shapes of the Germanium profile.

with the low field mobility in the emitter μ_{E0}. In case of an almost constant highly doped emitter, both μ_{nE0} and n_{iE} can be assumed as spatially independent. Therefore, the weight factor depends on the ratio of the low field mobilities and intrinsic carrier densities in the base and emitter only. The bias dependent values of h_{mE} are displayed in Fig. 2.41, showing fairly constant curves in the medium injection region. The weight factor is directly correlated to n_{ir}, which is largest for the box transistor.

Neutral base

Since in the neutral base $\nabla\psi = 0$, an increase of the electron density will also cause an increase of the holes. Furthermore, a potential drift field is assumed across the complete neutral base. Thus, inserting (2.29) and (2.33) into (2.71) can be used to derive an equation of the weight factor h_{mB}. A closed form solution was given in [PS14b]

$$h_{mB} = c_{hmB} \frac{(1-\theta)\exp\left(-\frac{x_e}{a_{ni}}\right) - a_{ni}\theta\left[\exp\left(-\frac{x_c}{a_{ni}}\right) - \exp\left(-\frac{x_e}{a_{ni}}\right)\right]}{(1-\theta)a_{ni}\left[\exp\left(\frac{x_c-x_e}{a_{ni}}\right) - 1\right] + \theta(x_c - x_e)}, \qquad (2.121)$$

with c_{hmB} similar to (2.96) only replacing μ_{nBE} with a mean mobility in the base μ_{nB} and

$$\theta = \frac{f_\zeta}{f_\zeta - \frac{\mu_{nBE}E_x}{v_{sn}} - 1}. \qquad (2.122)$$

For large changes of the bandgap and thus large electric fields, $(\mu_{nBE}E_x)/v_{sn}$ is in the order of -1. Therefore, $\theta = 1$ can be assumed in this case, allowing to simplify

(2.121) to the rather simple form

$$h_{\mathrm{mB}}(V_{\mathrm{B'E'}}, V_{\mathrm{B'C'}}) = h_{\mathrm{mB0}} \frac{\exp(u_{\mathrm{hBC}}) \left[\exp(u_{\mathrm{hmB}}) - 1 \right]}{u_{\mathrm{hmB}}}, \qquad (2.123)$$

with $u_{\mathrm{hmB}} = w_{\mathrm{B}}/a_{\mathrm{ni}}$, (2.110) and

$$h_{\mathrm{mB0}} = c_{\mathrm{hmB}} \exp\left(-\frac{x_{\mathrm{c0}}}{a_{\mathrm{ni}}}\right). \qquad (2.124)$$

Note the different sign of u_{hBC} due to the different coordinates x and x'. The application of the model is given in Fig. 2.42, showing excellent agreement in the relevant operating range. For small gradients in the Germanium profile, n_{i} can be assumed as constant, leading also to a constant weight factor.

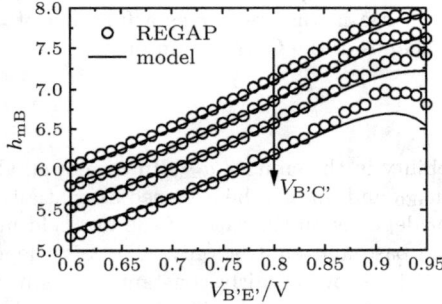

Figure 2.42: Modeling the weight factor for minorities in the neutral base using (2.123) for a transistor with a Germanium gradient across the complete base. Curves are given for $V_{\mathrm{B'C'}} = [-1.0, -0.5, 0, 0.5]\,\mathrm{V}$.

Since (2.123) is again similar to (2.97) and (2.106), similar models for the temperature dependences can be applied for h_{mB0} and u_{hmB}, though in a slightly more complicated form due to the additional exp-term in the numerator.

2.4.3 Compact model of the transfer current

Based on the equations derived in sec. 2.4.2, the compact model equation for I_{T} finally reads

$$I_{\mathrm{T}} = \frac{c_{10}}{Q_{\mathrm{p0}} + h_{\mathrm{jEi}}Q_{\mathrm{jEi}} + h_{\mathrm{jCi}}Q_{\mathrm{jCi}} + Q_{\mathrm{f,T}}}, \qquad (2.125)$$

with

$$Q_{\mathrm{f,T}} = h_{\mathrm{f0}}Q_{\mathrm{f0}} + h_{\mathrm{fE}}\Delta Q_{\mathrm{Ef}} + \Delta Q_{\mathrm{Bf}} + h_{\mathrm{fC}}Q_{\mathrm{fC}}, \qquad (2.126)$$

and is compatible with the transfer current formulation from HICUM/L2, with the addition of the weight factor h_{f0} and the bias dependence of h_{jEi} as well as the temperature dependence for all weight factors.

In the following chapters, the model equations for inclusion into the compact model HICUM/L2 are given. Visualizations of the compact model equations here are still given for 1D simulations results.

2.4.3.1 Low bias weighted charge

As low-bias weighted charge, the term

$$Q_{\mathrm{pT,low}} = Q_{\mathrm{p0}} + h_{\mathrm{jEi}} Q_{\mathrm{jEi}} + h_{\mathrm{jCi}} Q_{\mathrm{jCi}} \tag{2.127}$$

(cf. (2.125)) is considered. Here, the weight factors h_{jEi} and h_{jCi} are discussed with the equations (2.103) and (2.116). Two singularities exist for those equations. Next, only the methods to avoid these for h_{jEi} are explained. The smoothing of h_{jCi} is carried out in the same way.

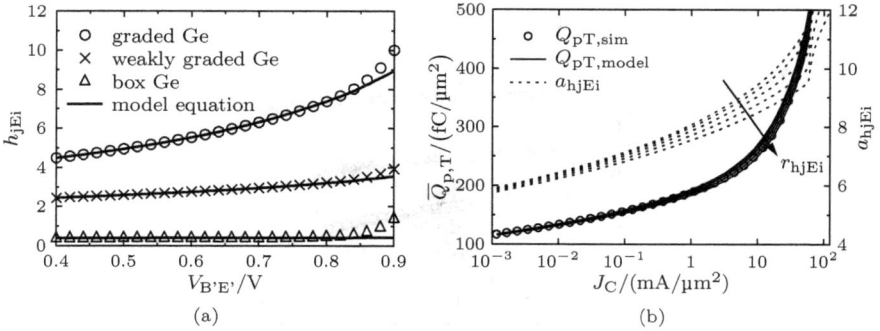

Figure 2.43: (a) Model for h_{jEi} calculated from terminal currents according to (2.90) for transistors with different doping profiles at $V_{\mathrm{B'C'}} = 0\,\mathrm{V}$. The increase of h_{jEi} from terminal currents is caused by the beginning influence of the weighted minority charge. (b) Smoothing of $Q_{\mathrm{p,T}}$ from low to medium current densities using r_{hjEi}. The value is swept here from $1\ldots5$.

The singularities exist for $V_{\mathrm{B'E'}} = 0\,\mathrm{V}$ and $V_{\mathrm{B'E'}} > V_{\mathrm{DEi}}$. Both are avoided by using dedicated smoothing functions. The smoothing for the upper boundary is performed by defining

$$v_{\mathrm{j,upp}} = V_{\mathrm{DEi}} - r_{\mathrm{hjEi}} V_{\mathrm{T}} \frac{x_{\mathrm{upp}} + \sqrt{x_{\mathrm{upp}}^2 + a_{\mathrm{fj}}}}{2}, \tag{2.128}$$

with

$$x_{\mathrm{upp}} = \frac{V_{\mathrm{DEi}} - V_{\mathrm{B'E'}}}{r_{\mathrm{hjEi}} V_{\mathrm{T}}}. \tag{2.129}$$

Next, the smoothing for the lower boundary is performed in a similar way by

$$v_{j,low} = V_T + V_T \frac{x_{low} + \sqrt{x_{low}^2 + a_{fj}}}{2},$$

(2.130)

with

$$x_{low} = \frac{v_{j,upp} - V_T}{V_T}.$$

(2.131)

Finally, $V_{B'E'}$ in (2.98) is replaced by $v_{j,low}$. Basically, the upper singularity could also be avoided by using C_{jEi0}/C_{jEi}, but applying (2.128) with (2.129) and introducing the additional model parameter r_{hjEi} allows a smooth transition between low and medium injection. The value of a_{fj} is fixed to 1.921 812 as already used in HICUM/L2. Note that above smoothing functions introduce additional temperature dependences due to V_T. However, a better agreement was obtained over a large range of temperatures when using V_T rather than a fixed value.

As shown in Fig. 2.43(a), very accurate results for modeling the weight factor can be obtained by above equations, which corresponds to an improved model for the normalized transconductance, cf. Fig. 2.44. The usage of r_{hjEi} to model the transition between low and medium injection is visualized in Fig. 2.43(b).

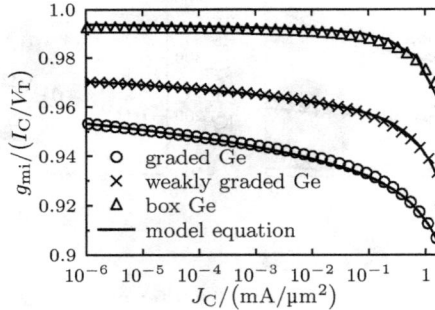

Figure 2.44: Normalized transconductance with modeling the weight factor h_{jEi} with (2.103) for transistors with different Germanium profiles at $V_{B'C'} = 0\,\text{V}$.

2.4.3.2 Weight factor for the low current mobile charge

In the GICCR formulation used in HICUM/L2, several components of the minority charge are summarized into

$$Q_{f0} = I_{Tf}\tau_{f0},$$

(2.132)

where τ_{f0} is only dependent on $V_{B'C'}$ ([SL99]). In terms of physical components, this charge is the sum of the charges from several transistor regions

$$Q_{f0} = Q_{mE} + f_{mBE}Q_{mBE} + Q_{mB} + \Delta Q_{BC}.$$

(2.133)

In this equation, f_{mBE} is close to 0. This is caused by common extraction practice. In HICUM/L2, no specific model for Q_{mBE} is implemented, although suitable approaches exist (e.g. [SC10]). The bias dependent transit time caused by this charge has a similar shape to that from Q_{jEi}, making it in fact very difficult (if not impossible) to separate both components experimentally. As a result, large portions of the transit time from Q_{mBE} are included in the extracted model for C_{jEi}. This is verified in Fig. 2.45(a). Here, the value from HICUM is calculated by using the previously extracted τ_{f0} and (2.132). As shown, only a small portion of Q_{mBE} is included in the modeled charge.

In contrast, the corresponding weighted charge reads

$$Q_{f0,T} = \frac{h_{mE}Q_{mE} + f_{mBE,T}h_{mBE}Q_{mBE} + h_{mB}Q_{mB} + h_{pBC}\Delta Q_{BC}}{h_{00}}, \qquad (2.134)$$

with the empirical parameter $f_{mBE,T} \leq 1$. Here, the weighted charge in the BE-SCR is included, which is due to the extraction method described in sec. 3.3.1. The extraction is generally only performed in the low-bias region where Q_{mBE} has no influence on the small-signal behavior. Thus, the parameters cannot predict the influence of $h_{mBE}Q_{mBE}$. As shown in Fig. 2.45(b), almost the complete charge is included in $Q_{f0,T}$, which is extracted for HICUM by using the extraction method in 3.3.1.

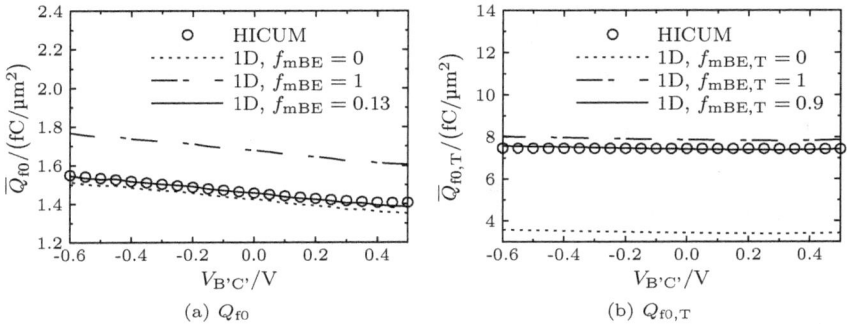

Figure 2.45: Comparison of Q_{f0} and $Q_{f0,T}$ from (2.133) and (2.134) with parameters extracted from the internal charges of 1D simulations and different values for f_{mBE} and $f_{mBE,T}$, respectively. Results are given for $V_{B'E'} = 0.9\,\text{V}$.

From (2.133) and (2.134) follows for the weight factor

$$h_{f0} = \frac{\Delta Q_{f,T}}{\tau_{f0}I_{Tf}} = \frac{1}{h_{00}}\frac{h_{mE}Q_{mE} + f_{mBE,T}h_{mBE}Q_{mBE} + h_{mB}Q_{mB} + h_{pBC}\Delta Q_{BC}}{Q_{mE} + f_{mBE}Q_{mBE} + Q_{mB} + \Delta Q_{BC}}, \qquad (2.135)$$

where the LHS of above equation is related to HICUM and the RHS corresponds to internal quantities. As shown in Fig. 2.46, this weight factor is not constant but depends on $V_{B'C'}$. However, in contrast to the model derived for h_{mB} (2.123), the value is increasing with $V_{B'C'}$ rather than decreasing.

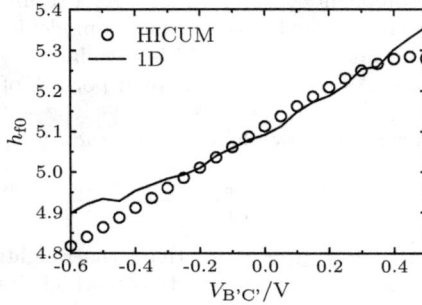

Figure 2.46: Weight factor h_{f0} extracted for HICUM and calculated from internal values with (2.135).

A general explanation valid for all types of transistor profiles cannot be given due to the large number of components included in h_{f0}. The decrease here is caused by two factors. First, the charge in the BC-SCR is decreasing with $V_{B'C'}$ due to decreasing width of the SCR. On the other hand, Q_{mB} is increasing. However, it increases to a smaller extend compared to ΔQ_{BC} due to the narrow neutral base. Further, Q_{mBE} and Q_{mE} are independent of $V_{B'C'}$ for a given $V_{B'E'}$ before the onset of high current effects, causing the effective charge to decrease, which is also shown in Fig. 2.45(a).

Due to the graded Germanium in this transistor, the weight factor h_{mBE} is much larger than h_{mB} and h_{mE} due to the smaller bandgap in both regions. It is also large compared to h_{pBC} due to the still larger electric field in the BC-SCR. h_{mBE}, though, is constant with $V_{B'C'}$, leading to an almost constant $Q_{f0,T}$. This effect is visualized in Fig. 2.45(b). As a consequence, the effective weight factor is increasing.

The voltage dependence with respect to $V_{B'E'}$ can be neglected in a production type compact model, because the $V_{B'E'}$ range where $Q_{f0,T}$ dominates the transfer current is quite small (approx. $[10 \ldots 15]$ mV). In this region, $Q_{f0,T}$ will not change significantly. Also, due to the strong current dependence of the charge in this region, an accurate extraction will be fairly difficult. On the other hand, the range of $V_{B'C'}$ can be quite high, including small forward and large reverse bias. This can be seen very well in the extraction results presented later in Fig. 3.12.

The relation of the bias dependence of Q_{mB} and Q_{pBC} strongly depends on the base and collector profile. Furthermore, the strong impact of $h_{mBE}Q_{mBE}$ is only given for transistors with graded Germanium profiles. Therefore, a general conclusion cannot be given here. A model for h_{f0} taking all possible bias dependences into account with physics-based parameters cannot be given in a compact form.

Thus, a semi-empirical model in the form

$$h_{f0}(V_{B'C'}) = h_{f00}\left(1 + a_{hf0c}V_{B'C'}\right) \tag{2.136}$$

is used here. The application shown in Fig. 2.47 justifies the use of a linear function for the $V_{B'C'}$ dependence.

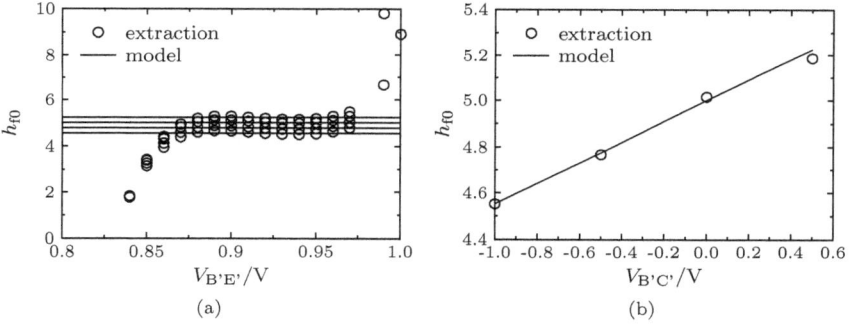

Figure 2.47: Application of (2.136) on values obtained from parameter extraction. Values for $h_{f0}(V_{B'C'})$ are given for $V_{B'E'} = 0.9\,\mathrm{V}$.

   In early releases of the model including h_{f0} (see [PSMK10]), only a constant value was included, which provided a very good agreement for different advanced SiGe technologies (e.g. [ARS$^+$10,PSK$^+$11,PSF13]).

   Although introducing the bias dependence of h_{f0} leads to additional parameters and additional extraction effort, this model is required to avoid a possible negative output conductance when applying $h_{f0} = $ constant. The reason behind this was discussed in [PS14a] and is caused by the $V_{B'C'}$-dependence of τ_{f0}. The effect is visualized in Fig. 2.48. Applying a constant h_{f0} leads to a change in the sign of g_o, which is not existing in the reference numerical simulations and also not the case with using the GICCR master equation based on internal data[1]. However, employing the (though semi-physical but still derived based on physical effects) model for $h_{f0}(V_{B'C'})$ avoids this error. Finally the parameter for the $V_{B'C'}$-dependence can be extracted fairly easy and accurate (cf. sec. 3.3.1).

2.4.3.3 Weight factors for the high current mobile charges

The weight factor for the high-current charge in the emitter follows directly from (2.120) and is given by

$$h_{fE} = \frac{\overline{\mu_{nB}n_{iB}^2}}{\mu_{nE}n_{iE}^2}. \tag{2.137}$$

---

[1]The deviations of the GICCR for the actual simulation results are caused by the numerical integration.

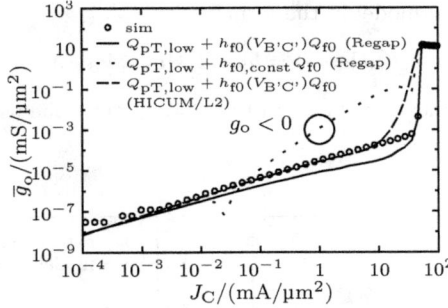

Figure 2.48: Modeling of the output conductance with and without a bias dependent h_{f0}. The picture is similar to what was published in [PS14a].

Similar, a model for the collector charge is given by

$$h_{\mathrm{fC}} = \overline{\frac{\mu_{\mathrm{nB}} n_{\mathrm{iB}}^2}{\mu_{\mathrm{nC}} n_{\mathrm{iC}}^2}}. \tag{2.138}$$

Both models are already given in [SC10]. For a heterojunction in the collector, $n_{\mathrm{iC}} \ll n_{\mathrm{iB}}$, leading to $h_{\mathrm{fC}} \gg 1$. This large weight factor can very accurately capture the strong impact of the Kirk-effect on the transfer current shown in sec. 2.4.1.5.

2.4.3.4 Temperature model

Due to the influence of $h_0(V_{\mathrm{B'E'}})$ and the thus altered temperature dependence, the equations for h_{jEi0} and a_{jEi} are defined in a more flexible way, giving

$$h_{\mathrm{jEi0}}(T) = h_{\mathrm{jEi0}}(T_0) \exp\left(\frac{\Delta V_{\mathrm{gBE}}}{V_{\mathrm{T}}} \left[\left(\frac{T}{T_0} \right)^{\zeta_{\mathrm{VgBE}}} - 1 \right] \right), \tag{2.139}$$

and

$$a_{\mathrm{hjEi}}(T) = a_{\mathrm{hjEi}}(T_0) \left(\frac{T}{T_0} \right)^{\zeta_{\mathrm{hjEi}}}, \tag{2.140}$$

with the model parameters ΔV_{gBE}, ζ_{VgBE} and ζ_{hjEi}. For h_{jCi}, the same equations are obtained.

The weight factor h_{f0} is modeled by

$$h_{\mathrm{f0}}(T) = h_{\mathrm{f0}}(T_0) \exp\left(\frac{\Delta V_{\mathrm{gBE}}}{V_{\mathrm{T}}} \left(\frac{T}{T_0} - 1 \right) \right). \tag{2.141}$$

The equation is based on the assumption that in the relevant operating range h_{f0} is dominated by the charge in and close to the BE-SCR Q_{mBE}. Therefore the temperature dependence is caused by the associated bandgap difference. This approach avoids the introduction of additional parameters. Application of above equations to

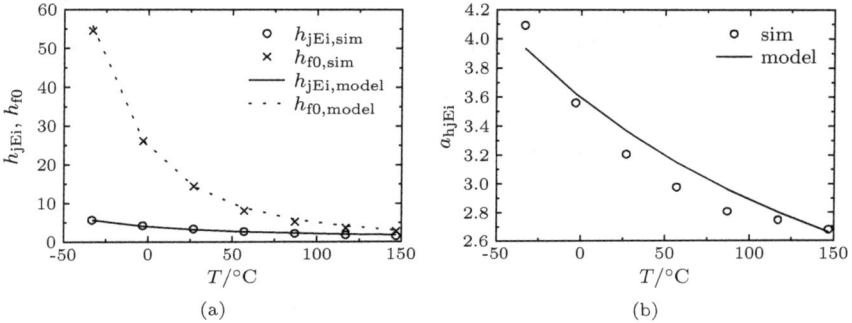

Figure 2.49: Application of (2.139), (2.140) and (2.141) to values obtained from extraction of numerical device simulations.

weight factors obtained from extraction are given in Fig. 2.49(a) and (b).

In the same way, the model equations for the temperature dependence of h_{fE} and h_{fC} are derived and read

$$h_{fE}(T) = h_{fE}(T_0) \exp\left(\frac{V_{gB} - V_{gE}}{V_T}\left(\frac{T}{T_0} - 1\right)\right), \tag{2.142}$$

and

$$h_{fC}(T) = h_{fC}(T_0) \exp\left(\frac{V_{gB} - V_{gC}}{V_T}\left(\frac{T}{T_0} - 1\right)\right), \tag{2.143}$$

respectively. The bandgaps, which are inserted in both equations are no new model parameters but the same values as obtained from the temperature models of the saturation currents. As shown in Fig. 2.50, the weight factors can be accurately described as a function of temperature by both equations.

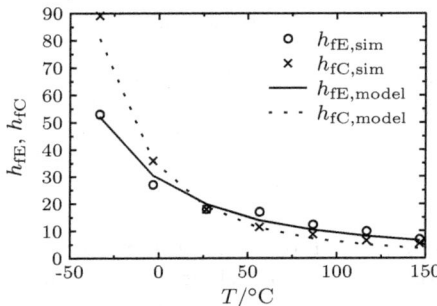

Figure 2.50: Application of the temperature models (2.142) and (2.143) on values obtained from extraction.

Applying equations (2.139)-(2.143) together with the equations from chapters 2.4.3.1-2.4.3.3 results in a very accurate model for the transfer current as a function of bias and temperature as shown in Fig. 2.51.

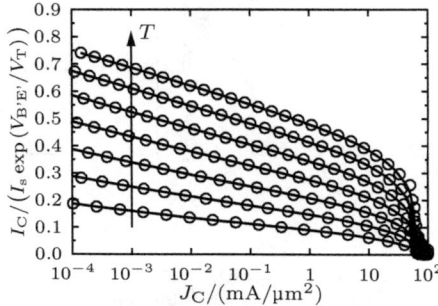

Figure 2.51: Normalized collector current for $T = [240\ldots420]\,$K for the strongly graded Germanium profile. Note that a different method for obtaining I_s (taking $h_{jEi}(V_{B'E'})$ correctly into account) was used here in comparison to Fig. 2.27, leading to slightly different values.

2.4.3.5 Discussion of different modeling approaches

In literature, a few different modeling approaches exist, which are discussed briefly in this section.

The model in [PKH01] was derived for a graded Germanium across the complete base with the same model for n_i as (2.33). Though the derived equation is different from the equations shown in this work, the explanation of the physical effect is the same. However, in [PKH01], the effects of both junctions cannot easily be distinguished as it is possible with the separate h_{jEi} and h_{jCi}. Furthermore, no discussion on the temperature dependence was included in the model, further limiting in practical relevance.

The authors of [PJ01] followed a different approach by modeling all effects of the transfer current in the corresponding non-ideality factor. In contrast to the model shown in this work, a more elaborate model for the doping profile in the base was used, leading to a fairly complicated model equation, though a less complicated formulation is given for the case of a uniformly doped base. Unfortunately, no discussion on parameter extraction and no verification of the derived equation are given in the publication, not even to results from numerical simulations.

In [HCS11a,HCS11b] a different model for the same effects as discussed here was published. The model was based on a completely different approach. Rather than modeling weight factors and charges separately, an effective hole concentration

$$N_{DC}(x) = h(x)N(x),\qquad(2.144)$$

with the actual net doping concentration N was defined. Since the equations of the depletion capacitance can be applied to all possible doping concentrations by adjusting the grading factor and the diffusion voltage, those equations were employed for the weighted charge. From the transfer current, "DC" parameters of the BE depletion capacitance are extracted. Despite this being a valid approach, the model is flawed because the location of x_e was still modeled with the electrostatic depletion capacitance obtained from S-parameter measurements. However, for a transistor with graded Germanium in the base, a "depletion capacitance" based on N_{DC} will have a completely different bias dependence than the capacitance from N. Thus, inconsistent results were obtained from this approach, often leading to very large values for the grading factor $z_{E,DC}$.

2.4.4 A comparative study of weight factors from DD and HD simulation

The limits of DD simulations have been discussed in chapter 2.2. Nevertheless, the GICCR, which is based on the DD transport equation, is still used as a starting point for the derivation of the bias and temperature dependences of the weight factors. In the following, a comparison between the weight factors derived from DD values and HD simulation is presented in order to show the applicability of the derived models, especially for transistors with strong non-local effects.

2.4.4.1 The weight function from HD simulations

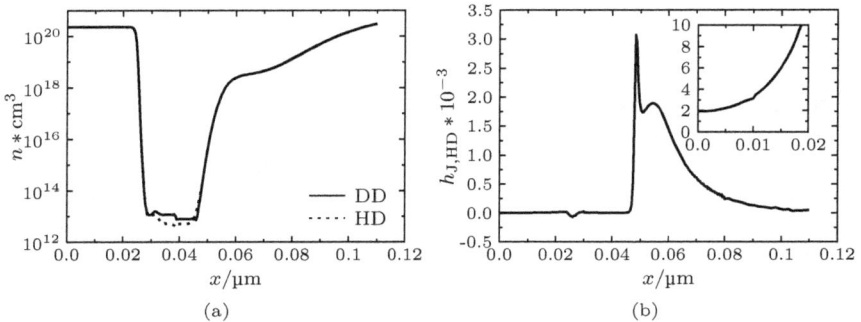

Figure 2.52: (a) Electron density of DD and HD simulations of a transistor with graded Germanium in the base for $V_{B'E'} = 0.6\,\text{V}$ and $V_{C'E'} = 1.0\,\text{V}$. (b) Spatial dependence of $h_{J,HD}$ for the same operating point. The inset is a zoom on the neutral emitter.

A derivation of the weight function for electron transport taking non-local effects into account was presented in [Paw08]. The basic idea behind it was splitting the electron current into a component with φ_n as the driving force and one with the gradient of the electron temperature. Using only the former component, the derivation was comparable to the DD formulation. The finally derived weight function contains five parts compared to the three parts from DD. The weight functions are

$$h_{g,\mathrm{HD}}(x) = \frac{\overline{\mu_{\mathrm{nr}} n_{\mathrm{ir}}^2}}{\mu_{\mathrm{x}}(x) n_{\mathrm{i,eff}}^2(x)} \quad , \quad h_{\mathrm{J,HD}}(x) = -\frac{A_{\mathrm{E}} J_{\mathrm{n},\varphi_n}(x)}{I_{\mathrm{T}}} \quad \text{and}$$

$$h_{v,\mathrm{HD}}(x) = \exp\left(\frac{V_{\mathrm{B'E'}} - \varphi_p(x)}{V_{\mathrm{T,L}}}\right). \tag{2.145}$$

For details regarding the components refer to [Paw08]. Although these three weight functions are similarly defined to (2.69), a notable difference exists for $h_{\mathrm{J,HD}}$, where not the complete current density but only the part corresponding to the gradient of φ_n is inserted. The two remaining HD weight functions are

$$h_{\psi,\mathrm{HD}}(x) = \exp\left(\frac{(V_{\mathrm{Tn}} - V_{\mathrm{Tp}})\psi}{V_{\mathrm{Tn}} V_{\mathrm{Tp}}}\right) \quad \text{and} \quad h_{\mathrm{T,HD}}(x) = \frac{V_{\mathrm{T,L}}}{V_{\mathrm{Tn}}}, \tag{2.146}$$

which basically take effects of the carrier energy into account. In above equations, V_{Tn} and V_{Tp} are the thermal voltage with respect to the electron and hole carrier temperatures while $V_{\mathrm{T,L}}$ is calculated from the lattice (i.e. ambient) temperature.

In operating points and regions of the transistor where $h_{\mathrm{J,HD}}$ is equal to h_{J} and thus close to one, the same weight function as for DD is obtained. Shown in Fig. 2.52 is the spatial dependence of $h_{\mathrm{J,HD}}$ for a low-bias operating point. Noticeable from this curve is that in the complete transistor region, including the neutral regions near the contacts, the value significantly differs from 1 and is even negative in certain regions, although the electron density is the same for DD and HD simulations. This effect is caused by the energy balance equation and the boundary conditions in HD simulations ([Wed]).

Though these large and potentially negative values finally lead to mathematically correct weight factors, no physically meaningful correlation of the weight factors with internal device quantities can be derived. Thus, the following comparison is based on weight factors extracted from terminal currents. Moreover, since simulations with lattice temperatures different from 300 K are not possible with the utilized simulator, the temperature models cannot be applied to HD simulation data.

2.4.4.2 Weight factors of the depletion charges

Both relevant weight factors, h_{jEi} and h_{jCi}, are extracted using the method explained in Ch. 3.3.1. However, as shown in the introduction of HD results (cf. sec. 2.2), output curves using HD equations are very questionable. Therefore, the evaluation of h_{jCi} is not given.

The extraction results for a_{hjEi} and the resulting curve for h_{jEi} along with the application of (2.103) are shown in Fig. 2.53. A larger value of a_{hjEi} is extracted for HD simulations, though with the same trend. Since in both cases, almost constant values are extracted, the model for h_{jEi} gives accurate results.

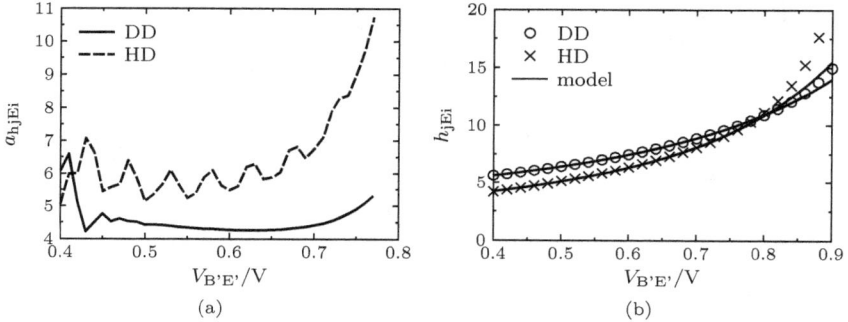

Figure 2.53: (a) Comparison of the extraction results for a_{hjEi} based on (3.15) for DD and HD simulations. (b) Application of (2.103) to the extracted curves of h_{jEi} for DD and HD simulations. Both pictures are generated for $V_{B'C'} = 0\,\text{V}$.

The model is still applicable, because in HD simulations the current is defined by the injected electrons in the neutral base and the velocity in the BC-SCR. Though the distribution is likely not a Boltzmann distribution, it still depends on the bandgap at the injection point. On the other hand, since the depletion capacitances and thus the location of x_e is the same for DD and HD ([Paw08]), the basic mechanisms, which led to (2.103), still exist in HD simulations.

2.4.4.3 Weight factors of the mobile charges

Here, only the weight factors h_{f0} and h_{fE} are shown in Fig. 2.54, but results are basically similar for h_{fC}. The general shapes of both weight factors for DD and HD is the same, except for increased bias dependences. However, the equation for h_{f0} is able to model both curves[1], while the $V_{B'C'}$ dependence of h_{fE} is caused by inaccuracies of the τ_{fE} model (cf. [Paw08]).

2.4.4.4 Summary

The previous sections have shown accurate models of weight factors extracted from HD simulations based on equations derived by the DD approach. Although all non-local effects are not taken into account by the model equations, the results are promising regarding the future application of the latter equations for next-generation high-speed transistors. In fact, published results (e.g. [PSF13] and [PSK$^+$09b, SKR$^+$11]) show the successful application of the model equations to both fabricated

---

[1]For DD results refer Fig. 2.47b.

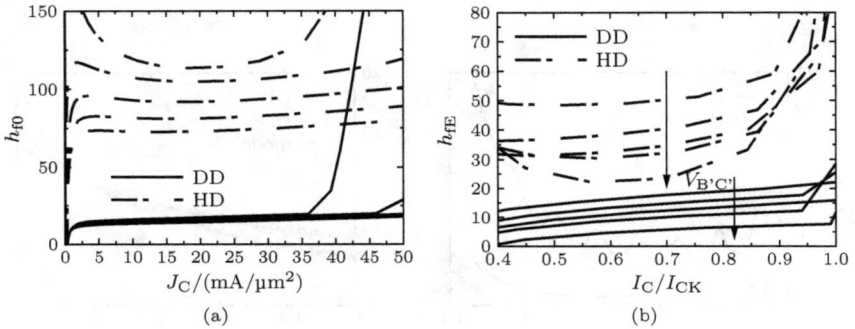

Figure 2.54: Comparison of the extraction results for h_{f0} and h_{fE} for $V_{B'C'} = [-1.5 \ldots 0.5]$ V.

transistors with a strong influence of non-local effects and anticipated transistors far in the future.

2.4.5 Application to experimental results

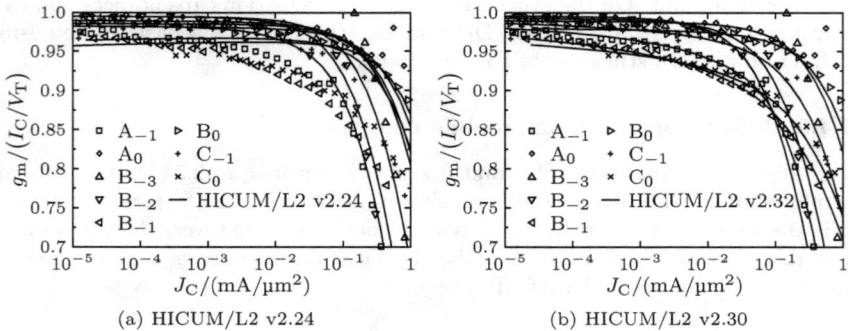

Figure 2.55: Application of the model for h_{jEi} and h_{f0} for transistors of different technologies. Pictures are taken from [PS14a].

The application of the derived model equations for a large number of technologies is given in [PS14a]. The results shown in Fig. 2.55 clearly demonstrate the improvements obtained when employing the introduced models. Parameters for different generations (numerical index) of different technologies (A, B, C)[1] are extracted for

---

[1]For more informations on the nomenclature of the technologies refer to [PS14a].

the existing HICUM/L2 v2.24 model. As shown in (a), the shape of the normalized transconductance cannot be modeled accurately with the model equations employed in HICUM/L2 v2.24. Since those equations are representative for all more advanced production type compact models, those results can be used as generalization.

However, the models discussed in this work were implemented in the HICUM/L2 v2.30 model (and all subsequent versions). Parameters extracted for this model cf. Fig. 2.55(b) provide excellent agreement for all selected technologies and process generations. Additional results are presented during the description of the extraction methodology (chapter 3.3) and the extraction results for the given technology (chapter 4.4) and also in [PSK09a] and [PSK+11].

2.5 Modeling of the electric field in the base-collector space charge region

The electric field at the BC-junction is one of the key components in the operation of bipolar transistors. Except for high injection effects, i.e. when the number of injected electrons in the neutral base exceeds the base doping, almost all relevant effects at high current densities are connected to the reduction of the electric field ([Kir62]). Thus, employing a model for the bias dependence of the electric field in a compact model allows to describe several effects in a compact model by physics-based equations.

One example is the charge ΔQ_{BC} from the previous chapter. In all presently available versions of HICUM/L2, this charge is implicitly modeled by the low current minority charge. Utilizing a model for the electric field would allow to correctly calculate this component as part of the depletion charge.

2.5.1 General model approach

Based on the Drift-Diffusion model two electric fields can be discussed, calculated either from the electro-static potential ψ or the quasi-Fermi potential φ_n. As long as the electric field is high enough to support a pure drift current, both are the same. However, once a significant portion of the current is caused by diffusion the values and shapes of the electric fields start to differ. Yet, both fields are of physical relevance: the charge model in the collector is based on ψ via the Poisson equation, whereas the electron velocity model requires the knowledge of φ_n. Since this chapter mainly focuses on the charge model and the corresponding small-signal values, the electric field is defined as $E = -\mathrm{d}\psi/\mathrm{d}x$.

Modeling the electric field in the collector of a BJT, especially the value of the peak electric field at the metallurgic BC-junction E_{jC}[1], provides an in-depth view on the effects inside the devices. By using E_{jC} the current dependent BC-capacitance and the resulting transit times can be modeled accurately.

From the Poisson equation follows for the charge in the collector of the transistor

$$Q_{BC} = -\varepsilon A_E E_{jC}. \tag{2.147}$$

Defining the internal voltages $V_{B'E'}$ and $V_{B'C'}$ as independent variables for Q_{BC}, the change of the charge is expressed as

$$\mathrm{d}Q_{BC} = \left.\frac{\partial Q_{BC}}{\partial V_{B'E'}}\right|_{V_{B'C'}} \mathrm{d}V_{B'E'} + \left.\frac{\partial Q_{BC}}{\partial V_{B'C'}}\right|_{V_{B'E'}} \mathrm{d}V_{B'C'}. \tag{2.148}$$

However, from a physics point of view it is more useful to model Q_{BC} as a function of I_T rather than $V_{B'E'}$. Above equation is thus rewritten as

$$\mathrm{d}Q_{BC} = \left.\frac{\partial Q_{BC}}{\partial I_T}\right|_{V_{B'C'}} \left.\frac{\partial I_T}{\partial V_{B'E'}}\right|_{V_{B'C'}} \mathrm{d}V_{B'E'} + \left.\frac{\partial Q_{BC}}{\partial V_{B'C'}}\right|_{V_{B'E'}} \mathrm{d}V_{B'C'}. \tag{2.149}$$

---

[1] E_{jC} is defined as positive, while the actual electric field has a negative value.

The former component is correlated to a transit time using

$$\left.\frac{\partial Q_{BC}}{\partial I_T}\right|_{V_{B'C'}} \left.\frac{\partial I_T}{\partial V_{B'E'}}\right|_{V_{B'C'}} = \tau_{Cce} g_{m,BC}, \tag{2.150}$$

with

$$\tau_{Cce} = - \varepsilon A_E \left.\frac{dE_{jC}(V_{B'C'}, I_T)}{dI_T}\right|_{V_{B'C'}}. \tag{2.151}$$

The latter component in (2.149) corresponds to the current dependent BC-depletion capacitance and is expressed as

$$C_{jCi} = - \varepsilon A_E \left.\frac{dE_{jC}(V_{B'C'}, I_T)}{dV_{B'C'}}\right|_{V_{B'E'}}. \tag{2.152}$$

It is important to realize that τ_{CCe} is not a transit time correlated to minority charges, since it is caused by electrons in the collector and holes in the base. However, in the medium current region the τ_{CCe} has a similar shape to τ_{f0} (cf. Fig. 2.56). Thus, it is often modeled as a part of the total minority charge. As a consequence, employing a physics-based model for E_{jC} which accurately describes the dependence on I_T and $V_{B'C'}$ results in correct values for both components.

In the following, the general evaluation of the model is performed using two Si-BJTs representing a high-speed (HS) and a high-voltage (HV) profile, respectively. Silicon based transistors rather than SiGe HBTs are used to avoid non-ideal influences of the conduction band barrier. The characteristic data of the internal collector for both transistors is summarized are Tab. 2.2.

Parameter	HS	HV
N_{Ci}/cm^{-3}	10^{17}	10^{17}
$w_{Ci}/\mu\text{m}$	0.2	1.0
$V_{\lim}/\text{V}$	0.3	1.5
V_{PT}/V (2.165)	3.1	77

Table 2.2: Characteristic values for the transistors used for evaluating the electric field in this work.

The current dependence of the transit times for the HS transistor according to (2.151) is shown in Fig. 2.56 for a large range of $V_{B'C'}$. I_{CK} is defined as the current where $E_{jC} = E_{\lim}$ (for large $V_{C'E'}$) or the field approaches a flat shape in the collector (for small $V_{C'E'}$)[1]. Especially the normalized plot in Fig. 2.56(b) illustrates the large error resulting from the neglected field dependence in the standard minority transit time equations.

The general model of the electric field in the collector of a BJT is based on the

---

[1]For a short discussion on I_{CK} extracted from transit times compared to this definition refer to 2.5.1.4

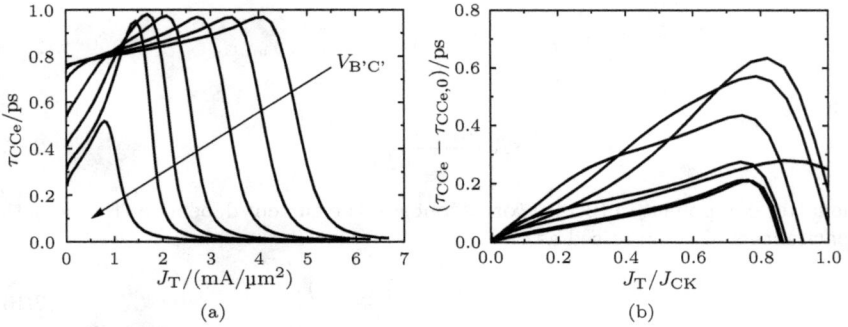

Figure 2.56: Current dependence of the transit times caused by the changing electric field for $V_{B'C'} = [-6.0\ldots0.6]\,\mathrm{V}$. On the right side the current is normalized to I_{CK} and only the increase of the transit time with respect to $\tau_{CCe,0} = \tau_{CCe}|_{I_T\to0}$ is shown in order to be in compliance with existing equations in HICUM/L2. Note that in the normalized plot no clear trend with $V_{B'C'}$ exists.

Poisson equation which for low injection in the BC-SCR simplifies to

$$\frac{\mathrm{d}E}{\mathrm{d}x} = \frac{q}{\varepsilon}(N_{Ci} - n)\ ,\qquad(2.153)$$

where N_{Ci} is the doping concentration in the collector and n is the electron density in the SCR. Assuming a pure drift current in the collector the electron density can be replaced by a function of the transfer current. Therefore, after introducing the transport equation for electrons eq. (2.153) changes to

$$\frac{\mathrm{d}E}{\mathrm{d}x} = \frac{q}{\varepsilon}\left(N_{Ci} - \frac{I_T}{q\mu_n E(x)}\right)\ .\qquad(2.154)$$

In classical theory, $\mu_n E(x)$ is often replaced by the saturation velocity of the electrons v_{sn}. However, this is only justified for large electric fields, i.e. $E(x) \gg E_{lim}$. In general v_n is a function of the electric field. Therefore, a differential equation has to be solved for the calculation of the electric field. The solution of the differential equation strongly depends on the electron velocity model in eq. (2.154). The boundary condition for this differential equation is derived by integrating the electric field across the complete collector and reads

$$\int_0^{w_{Ci}} E\,\mathrm{d}x = -V_{Ci}\ .\qquad(2.155)$$

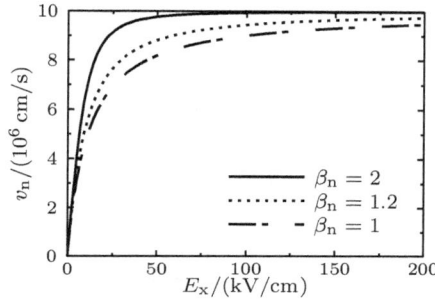

Figure 2.57: Application of (2.157) with different values of β_n. For $\beta_n = 1$ (2.157) simplifies to (2.158).

The collector voltage V_{Ci} is often calculated as

$$V_{Ci} = V_{DCi} - V_{B'C'}, \qquad (2.156)$$

which is only correct in the case of $N_{Ci} \ll N_B$. For highly doped collectors the SCR also extends into the base leading to a significant voltage drop there.

2.5.1.1 Mobility model

A model for the mobility often found in literature is (e.g. [CT67])

$$\mu_n = \frac{\mu_{n0}}{\left(1 + \left(\frac{|E|}{E_{lim}}\right)^{\beta_n}\right)^{1/\beta_n}}, \qquad (2.157)$$

with the doping, mole-fraction and temperature dependent low field electron mobility μ_{n0}, $E_{lim} = v_{sn}/\mu_{n0}$ and $\beta_n = 2$. Unfortunately, inserting (2.157) into (2.154) leads to a differential equation which cannot be solved analytically.

In [KH75] the field dependence of the electron velocity was modeled by a simpler function

$$v_n = v_{sn} \frac{E}{E_{lim} + E}, \qquad (2.158)$$

corresponding to (2.157) and $\beta_n = 1$. Although the shape of the mobility and its value around E_{lim} strongly depends on β_n and, thus, strongly differs when using $\beta_n = 1$ instead of 2, published values of β_n after adjusting the mobility to data obtained from BTE simulation are closer to 1 ([WS10]). Note that the published results rely on a driving force originating from the HD transport equation. However, similar results can be obtained for the pure DD model. Fig. 2.57 shows the comparison of the application of (2.157) with different values of β_n taken from literature.

The solution of the differential equation after inserting (2.158) into (2.154) reads

$$\frac{E_{jC} - E(x)}{a} - \frac{b}{a^2} \ln\left(\frac{aE(x) - b}{aE_{jC} - b}\right) = x, \tag{2.159}$$

with

$$a = \frac{qN_{Ci}}{\varepsilon}\left(1 - \frac{I_T}{I_{lim}}\right) \quad \text{and} \quad b = \frac{qN_{Ci}}{\varepsilon}\frac{I_T}{I_{lim}}E_{lim}. \tag{2.160}$$

and $I_{lim} = qA_E N_{Ci} v_{sn}$.

Although this is a closed-form solution of the electric field, it is not suited for compact models for several reasons. As discussed in [SC10] the numerical cost of calculating the bias dependent value of E_{jC} is too high. Also, it is only valid for $E(x) \neq b/a$ which corresponds to a horizontal field. Finally, the assumptions made in (2.154) limit the application to low injection, i.e. $n \ll N_{Ci}$. Thus, the application is also limited to the depletion zone. Hence, the correct width of the SCR has to be known for a partially depleted collector.

2.5.1.2 Operating ranges

For deriving a compact model the curves of $E_{jC}(V_{B'C'}, I_T)$ are separated into physics-based operating ranges. For this purpose it is useful to define

$$E_{CKl} = \frac{V_{Ci}}{w_{Ci}}, \tag{2.161}$$

which corresponds to the constant horizontal field. The separation of the operating regions depends on whether E_{CKl} is larger (*high-voltage case*) or smaller (*low-voltage case*) than E_{lim}. For practical high-speed SiGe HBTs with a high collector doping $V_{lim} = E_{lim}w_{Ci}$ is in the region of 0.2 to 0.5 V. Hence, the high-voltage case is most relevant for those transistors. On the other hand, the low-voltage case is of practical use for high-voltage transistors widely used in power amplifiers.

2.5.1.3 Low-current field

The low-current electric field E_{jC0}, which depends on $V_{B'C'}$ only, can be directly calculated based on the depletion charge. This charge follows from the extracted depletion capacitance. Using the field E_{jC00} at low current and $V_{B'C'} = 0\,\text{V}$, it follows

$$E_{jC0}(V_{B'C'}) = E_{jC00} - \frac{Q_{jCi}(V_{B'C'})}{\varepsilon A_E}. \tag{2.162}$$

For the calculation of E_{jC00} it is assumed that the junction is completely undepleted for $V_{B'C'} = V_{DCi}$, i.e. $V_{Ci} = 0\,\text{V}$. Thus, from (2.162) follows directly

$$E_{jC00} = \frac{Q_{jCi}(V_{DCi})}{\varepsilon A_E}. \tag{2.163}$$

The application of the latter two equations using the parameters extracted from

C_{jCi} without any further fine-tuning is shown in Fig. 2.58. Except for small deviations at large reverse bias a very good agreement is obtained. Note that for the HS and HV transistor $E_{\mathrm{jC}}(V_{\mathrm{Ci}})$ is almost identical for $V_{\mathrm{Ci}} < V_{\mathrm{PT}}$ due to the same doping in the base and collector (cf. Tab. 2.2).

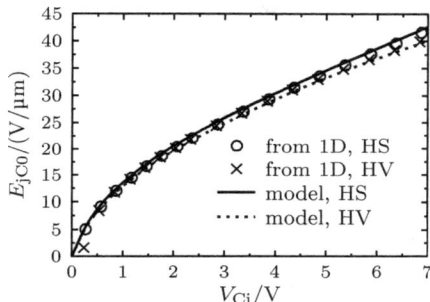

Figure 2.58: Modeling the bias dependence of E_{jC0} with (2.162) for the HS and HV transistor from Tab. 2.2. E_{jC00} was calculated using (2.163).

2.5.1.4 Onset of high-current effects

The onset current of high-current effects, I_{CK}, is defined differently for the low- and high voltage-case ([SC10]). For low-voltages the horizontal field ($E_{\mathrm{jC}} = E_{\mathrm{CKl}}$) defines I_{CK}. For high voltages $E_{\mathrm{jC}} = E_{\mathrm{lim}}$ is used. Since both definitions are based on E_{jC}, I_{CK} extracted from terminal characteristics is a key value for modeling the electric field at high currents. For instance, in [ST04] I_{CK} is used together with E_{lim} to define the bias dependence of E_{jC} starting from E_{jC0}. Therefore, the extraction of the model parameters for I_{CK}, most of which are re-used in the model presented in the following, is based on the transit time rather than on E_{jC}. However, the transit time increase $\Delta\tau_{\mathrm{fh}}$, which is the increase of τ_{fh} with respect to its low current value, is not necessarily the same at $E_{\mathrm{jC}} = E_{\mathrm{lim}}$ for different voltages. Thus, the extracted I_{CK} may differ from the one at $E_{\mathrm{jC}} = E_{\mathrm{lim}}$.

Sample results for both transistors are given in Fig. 2.59. For the HS transistor I_{CK} from the theoretical definition differs from the actually extracted value. However, the overall error is smaller than $10\,\%$ in the investigated bias region, causing only minor deviations when included into a model for E_{jC}.

2.5.2 High-voltage case

2.5.2.1 Low injection

This operation region is characterized by $E_{\mathrm{jC}} \geq E_{\mathrm{CKl}}$. Using (2.158) this corresponds to a current

$$I_{\mathrm{CKl}} = I_{\mathrm{lim}} \frac{E_{\mathrm{CKl}}}{E_{\mathrm{lim}} + E_{\mathrm{CKl}}}. \tag{2.164}$$

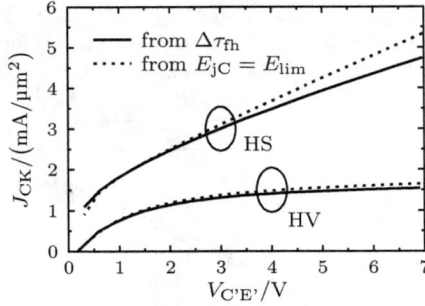

Figure 2.59: I_{CK} extracted for both transistors according to Tab. 2.2 from $\Delta\tau_{fh}$ and $E_{jC} = E_{lim}$.

In the classical theory I_{lim} is used as the current resulting in a horizontal field. However, the field dependent velocity can become significantly smaller than v_{sn} especially for E_{CKl} close to E_{lim}. Thus, I_{CKl} is a better indicator for the horizontal field. For large collector voltages though, $I_{CKl} \approx I_{lim}$.

Depending on V_{Ci} the collector is either completely or only partially depleted. In the latter case an ohmic region exists between the end of the SCR and the buried layer. The voltage dividing both regions is

$$V_{PT} = \frac{qN_{Ci}}{2\varepsilon} w_{Ci}^2.$$ (2.165)

However, in a certain bias range the partially depleted collector can still become fully depleted for higher currents. The spatial dependence of the electric field calculated from ψ is given in Fig. 2.60, showing both a fully and partially depleted collector. For the latter the collector becomes fully depleted for higher currents.

Fully depleted collector

According to the classical theory the electric field at the BC junction in this operating range is given by the linear function ([SC10])

$$E_{jC} = \frac{V_{Ci} + V_{PT}\left(1 - \frac{I_T}{I_{lim}}\right)}{w_{Ci}} = E_{jC0} - \frac{V_{PT}}{w_{Ci}}\frac{I_T}{I_{lim}}.$$ (2.166)

In this equation E_{CKl} is obtained for $I_T = I_{lim}$. This leads to larger errors around $E_{CKl} \approx E_{lim}$, i.e. at the transition to the low-voltage case. Thus, (2.164) is inserted instead of I_{lim} leading to

$$E_{jC} = E_{jC0} - (E_{jC0} - E_{CKl})\frac{I_T}{I_{CKl}}.$$ (2.167)

Above linear equation results in a constant transit time. However, the compari-

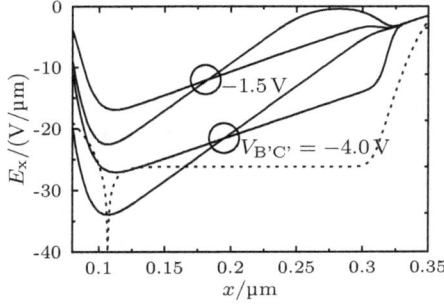

Figure 2.60: Spatial dependence of the electric field in the collector for two different $V_{B'C'}$ and $J_C = [0,1] \, \text{mA}/\mu\text{m}^2$. For $V_{B'C'} = -1.5 \, \text{V}$, no punch-through has occurred for low currents. However, the collector becomes fully depleted with increasing currents.

son to 1D device simulations shows an increase of the latter. In [Paw11], a method was presented to take this current dependence into account by multiplying I_{CKl} with a velocity dependent term. The application towards more advanced profiles has shown an undercompensation by this correction term. An additional source of error is introduced by the uncertainties in calculating V_{Ci} and E_{CKl}.

In the model proposed here, model equations using transit times are employed. The basic idea is that, similar to charges, also the electric field cannot be measured directly from transistor characteristics. Only its small-signal values, i.e. the derivatives according to (2.151) and (2.152), can be obtained from S-parameter measurements.

The model equation for the punch-through case is given by

$$E_{jC} = E_{jC0} - \frac{\tau_{CCe,0} I_{CKl}}{A_E \varepsilon} \left[\frac{I_T}{I_{CKl}} - \left(\frac{I_T}{I_{CKl}} \right)^{\zeta_{PT}} \right] + \left(E_{CKl} - E_{jC0} \right) \left(\frac{I_T}{I_{CKl}} \right)^{\zeta_{PT}} , \quad (2.168)$$

taking the known physical quantities into account. For $I_T \to 0$ follows $E_{jC,PT} = E_{jC0}$, while for $I_T = I_{CKl}$ $E_{jC,PT} = E_{CKl}$ is inserted. Furthermore, for $I_T \to 0$ follows $dE_{jC,PT}/dI_T = \tau_{CCe,0}$, linking the equation to measured transit times. The parameter ζ_{PT} with the default value 2 provides a further degree of freedom for fitting the transit times. For $\zeta_{PT} = 1$ the original formulation (2.167) is obtained.

Partially depleted collector

In this case a part of the collector is neutral and acting as an ohmic resistance. Thus, the shape of $E_{jC}(I_T)$ is different to the punch-through case. As discussed before even if the collector is not completely depleted at low currents, punch-through may still occur for increasing currents which are smaller than I_{CKl}. This effect was shown in Fig. 2.60. While for low currents the field E_{wC} at the end of the collector follows the field calculated for an ohmic region, it starts deviating beyond a specific current

I_{PT}. In this case E_{jC} will follow an almost linear curve (similar to (2.167)) which results in a different shape for τ_{Cce} after punch-through.

For modeling purposes the current $I_{PT} = f(V_{B'C'})$, which is the current defining the transition from punch-through to a partially depleted collector for a given voltage $V_{Ci} < V_{PT}$, is required to be known. This current is approximated by equating E_{wC} for the ohmic collector with E_{wC} for the punch-through case. The latter is given by

$$E_{wC} = \frac{V_{Ci}}{w_{Ci}} + \frac{V_{PT}}{w_{Ci}} \left(\frac{I_T - I_{CKl}}{I_{lim}} \right), \tag{2.169}$$

and uses $E_{wC} = E_{CKl} = V_{Ci}/w_{Ci}$ at $I_T = I_{CKl}$ rather than at I_{lim}. Based on (2.157) the current dependent electric field in the ohmic region is calculated as

$$E_{wC} = E_{lim} \frac{I_T}{I_{lim}} \left[1 - \left(\frac{I_T}{I_{lim}} \right)^{\beta_n} \right]^{-1/\beta_n}. \tag{2.170}$$

Calculating the intercept leads to

$$E_{lim} \frac{I_{PT}/I_{Ilim}}{1 - I_{PT}/I_{lim}} = E_{CKl} + \frac{V_{PT}}{w_{Ci}} \frac{I_{PT} - I_{CKl,\beta_n=1}}{I_{lim}}, \tag{2.171}$$

where β_n is set to 1 to enable an analytical solution. This solution reads

$$I_{PT} = I_{lim} \left(k - \sqrt{k^2 - \frac{I_{CKl,\beta_n=1}}{I_{lim}} + \frac{V_{Ci}}{V_{PT}}} \right), \tag{2.172}$$

with

$$k = \frac{I_{lim} + I_{CKl,\beta_n=1}}{2I_{lim}} - \frac{V_{Ci} + V_{lim}}{2V_{PT}}, \tag{2.173}$$

and $I_{CKl,\beta_n=1}$ as current calculated with $\beta_n = 1$ rather than using the extracted value. The application of the model is given in Fig. 2.61 showing a good agreement. The deviations are mainly caused by setting $\beta_n = 1$. Although by fine-tuning of the parameters E_{lim} and V_{PT} a better fit is possible, this would introduce two more parameters and is not recommended. The shape of I_{PT} in comparison to I_{CKl} can be explained as follows. For low-voltages both currents are the same. In this bias range no punch-through happens below I_{CKl}. For higher voltages I_{PT} decreases. Thus, punch-through happens at currents smaller than I_{CKl}, i.e. before the field becomes horizontal. For even higher voltages the collector is completely depleted even for very low injection as indicated by $I_{PT} < 0$.

For the partially depleted collector, E_{jC} calculated by ([SC10])[1]

$$E_{jC} = E_{wC} + E_{jC0} \sqrt{\left(1 - \frac{I_T}{I_{lim}} \right) \left(1 - \frac{E_{wC}}{E_{CKl}} \right)}. \tag{2.174}$$

Although the general trend of the field can be modeled by this equation, especially

---

[1]Some variables have been replaced by the nomenclature of this work

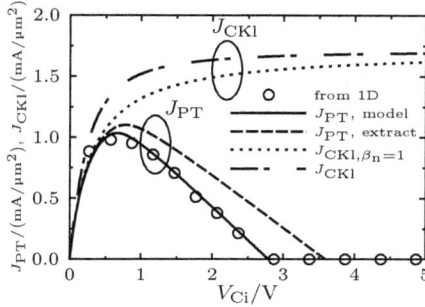

Figure 2.61: I_{PT} according to (2.172) normalized to $A_E = 1\,\mu m^2$. Shown are also the curves for the normalized values of I_{CKl} and $I_{CKl,\beta_n=1}$ of the simulated transistor. Note that the numerically more stable formulation (A.51) is used. In this plot "J_{PT}, model" represents (2.172) with physics-based parameters, particularly V_{PT} from (2.165). "J_{PT}, extract" uses the same equation but V_{PT} from the extraction of $I_{CK} = f(V_{CEi})$ based on $\Delta\tau_{fh}$.

the results for the transit time are too inaccurate. A more flexible option is using (2.174) only to calculate $E_{jC,PT}$ at I_{PT} and utilizing (2.168) with I_{CKl} replaced by I_{PT} and E_{CKl} replaced with $E_{jC,PT}$. The equation therefore reads

$$E_{jC} = E_{jC0} - \frac{\tau_{CCe,0} I_{PT}}{A_E \varepsilon} \left[\frac{I_T}{I_{PT}} - \left(\frac{I_T}{I_{PT}} \right)^{\zeta_{Ohm}} \right] - \left(E_{jC0} - E_{jC,PT} \right) \left(\frac{I_T}{I_{PT}} \right)^{\zeta_{Ohm}},$$
(2.175)

with a different model parameter ζ_{Ohm} (also close to 2). However, the accuracy of the modeled transit time strongly depends on very accurate values of $E_{jC,PT}$ which cannot be extracted from terminal quantities. Also, it would require to add several new parameters starting with the already discussed E_{lim} and V_{PT} separately from I_{CK} in (2.172). Small errors in either I_{PT} or $E_{jC,PT}$ already lead to significant deviations of $\tau_{CCe,PT}$ which is defined as $\tau_{CCe}(I_{PT})$.

As a consequence, a separate model for $\tau_{CCe,PT}$ is used in this work. A semi-empirical equation based on (2.172) is employed, reading

$$\tau_{CCe,PT} = \tau_{CCe,0} + \Delta\tau_{PT} \frac{I_{PT}}{I_{lim}},$$
(2.176)

with the model parameter $\Delta\tau_{PT}$. The equation assumes that the slope of τ_{CCe} for the ohmic case is independent of V_{Ci}. Thus, the transit time increase is modeled as a linear function of I_{PT}. The model is visualized in Fig. 2.62. It enables a good fit to the extracted transit time. In this figure, the reference 1D results for $\tau_{CCe,PT}$ where obtained from (2.151) at I_{PT}.

(2.175) is adopted using $\tau_{CCe,PT}$ based on (2.176) rather than $E_{jC,PT}$ and finally

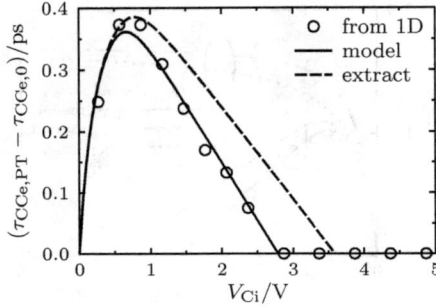

Figure 2.62: $\tau_{CCe,PT}$ according to (2.176) as a function of V_{Ci}. "model" and "extract" are defined similarly to Fig. 2.61.

reads

$$E_{jC,Ohm} = E_{jC0} - \frac{\tau_{CCe,0} I_{PT}}{A_E \varepsilon} \frac{I_T}{I_{PT}} - \frac{\Delta \tau_{PT} I_{PT}^2}{A_E \varepsilon \zeta_{Ohm} I_{lim}} \left(\frac{I_T}{I_{PT}} \right)^{\zeta_{Ohm}}. \tag{2.177}$$

The value for $E_{jC,PT}$ follows directly from inserting $I_T = I_{PT}$. For $I_{PT} < I_T < I_{CKl}$, the equation for the punch-through case is used but with I_{PT} as lower limit. It reads

$$E_{jC} = E_{jC,PT} - \frac{\tau_{CCe,PT}(I_{CKl} - I_{PT})}{A_E \varepsilon} \left[\frac{I_T - I_{PT}}{I_{CKl} - I_{PT}} - \left(\frac{I_T - I_{PT}}{I_{CKl} - I_{PT}} \right)^{\zeta_{PT}} \right]$$
$$- \left(E_{CKl} - E_{jC,PT} \right) \left(\frac{I_T - I_{PT}}{I_{CKl} - I_{PT}} \right)^{\zeta_{PT}}, \tag{2.178}$$

with the same parameter ζ_{PT} as in (2.168).

2.5.2.2 Medium injection and high injection

The current range of $I_{CKl} < I_T < I_{CK}$ is defined as medium injection. The same smoothing function as in [ST04] is employed, reading

$$E_{jC,med} = E_{inf} + \frac{e_j + \sqrt{e_j + g_{jC} E_{lim}^2}}{2}, \tag{2.179}$$

with $E_{inf} = V_T / w_i$ ([SC10]). The equation (2.180) for the field is adopted from the punch-through case (2.178) by replacing the lower boundary by I_{CKl} and E_{CKl} and the upper boundary by I_{CK} and E_{CK}

$$e_j = E_{CKl} - \frac{\tau_{CCe,PT}(I_{CK} - I_{CKl})}{A_E \varepsilon} \left[\frac{I_T - I_{CKl}}{I_{CK} - I_{CKl}} - \left(\frac{I_T - I_{CKl}}{I_{CK} - I_{CKl}} \right)^{\zeta_{PT}} \right]$$
$$+ \left(E_{CK} - E_{CKl} \right) \left(\frac{I_T - I_{CKl}}{I_{CK} - I_{CKl}} \right)^{\zeta_{PT}}, \tag{2.180}$$

with [ST04]

$$E_{CK} = E_{lim} \frac{V_{Ci}/V_{lim}}{\left(1 + \left(V_{Ci}/V_{lim}\right)^{\beta_n}\right)^{1/\beta_n}}. \tag{2.181}$$

In the original publication $\beta_n = 2$ was used. In this work rather than using β_n from the I_{CK} extraction a fixed value of 1 is inserted for two reasons. First, the term using $\beta_n = 1$ is already required in (2.172) and can thus be reused. Second, the impact of deviations can generally be neglected in the corresponding operating region.

2.5.3 Low-voltage case

Although it is theoretically possible that punch-through happens in the low-voltage case, it is not of practical interest here. Therefore, the equations of the partially depleted collector are employed. Especially using (2.181) rather than E_{lim} as characteristic high current field is crucial for applying the equations in the low-voltage case. As long as $I_{PT} = I_{CKl}$ holds no additional equations for I_{PT} and $\tau_{CCe,PT}$ are required. However, due to the small difference between I_{PT} and I_{CKl} numerical issues for the punch-through equations have to be avoided (cf. 2.5.5).

2.5.4 Low-current transit time

The low-current transit time $\tau_{CCe,0}$ is required for the modeling of the electric field in the low-injection case according to (2.168) and (2.177). A model is derived based on the simple physics-based equations for the punch-through and ohmic case given in (2.166) and (2.174), respectively.

For the ohmic case

$$\tau_{CCe,0} = \tau_{CCe,00} + \Delta\tau_{CCe,0}\left(c - 1\right) + \Delta\tau_{CCe,0}\frac{V_{lim}}{V_{DCi}}\left(\frac{V_{DCi}}{V_{Ci}}c - 1\right) \tag{2.182}$$

is derived. Here, $\tau_{CCe,00}$ is the transit time for $V_{B'C'} = 0\,\mathrm{V}$

$$\tau_{CCe,00} = \frac{A_E \varepsilon E_{jC00}}{2I_{lim}}\left(1 + \frac{V_{lim}}{V_{DCi}}\right) - \frac{\varepsilon V_{lim}}{w_{Ci}I_{lim}}, \tag{2.183}$$

and $\Delta\tau_{CCe,0}$ the parameter for describing the bias dependence given by

$$\Delta\tau_{CCe,0} = \frac{A_E \varepsilon E_{jC00}}{2I_{lim}}. \tag{2.184}$$

The normalized depletion capacitance c is defined as $c = C_{jCi0}/C_{jCi}$. In this equation C_{jCi} is inserted because the effect is correlated to the collector instead of the base. Therefore, $C_{jCi,t}$, i.e. the capacitance neglecting the punch-through effect, as utilized in the τ_0 model from HICUM/L2 cannot be used. Contrary to the ohmic case a constant transit time follows from (2.166) for the punch-through case as

$$\tau_{CCe,0} = \frac{A_E \varepsilon V_{PT}}{w_{Ci}I_{lim}}, \tag{2.185}$$

which can also be derived from (2.182) when inserting $V_{Ci} = V_{PT}$ and $C_{jCi} = C_{jCi}(V_{PT}) = C_{jCi,PT}$. Assuming $V_{DCi} < V_{PT}$, in this case the ratio c can be expressed by

$$c = \frac{C_{jCi0}}{C_{jCi,PT}} = \frac{w_{Ci}}{w_{BC,0}} = \sqrt{\frac{V_{PT}}{V_{DCi}}}, \tag{2.186}$$

where $w_{BC,0}$ is the width of the SCR at $V_{B'C'} = 0\,V$. Furthermore, from simple integration follows

$$E_{jC00} = \frac{2V_{DCi}}{w_{BC,0}} = \frac{2\sqrt{V_{DCi}V_{PT}}}{w_{Ci}}. \tag{2.187}$$

Inserting both equations (2.186) and (2.187) into (2.182) finally also leads to (2.185).

In terms of compact modeling eq. (2.182) has minor drawbacks. Compared to the existing equation in HICUM/L2 an additional term exists. The latter has uncertainties due its dependence on several parameters (assuming that $\Delta\tau_{CCe,0}V_{lim}/V_{DCi}$ is not merged into an additional parameter). Also, this term requires that V_{Ci} is limited to V_{PT}. The influence of the required limit of V_{Ci} is shown in Fig. 2.63. Deviations at large bias exist when using V_{Ci} directly. A simplified model using only the first two terms of (2.182) shows acceptable results only for the high speed device, although with errors at small voltages. However, for the high-voltage device the simplification cannot provide the correct shape. Therefore, the complete model is employed in the final implementation.

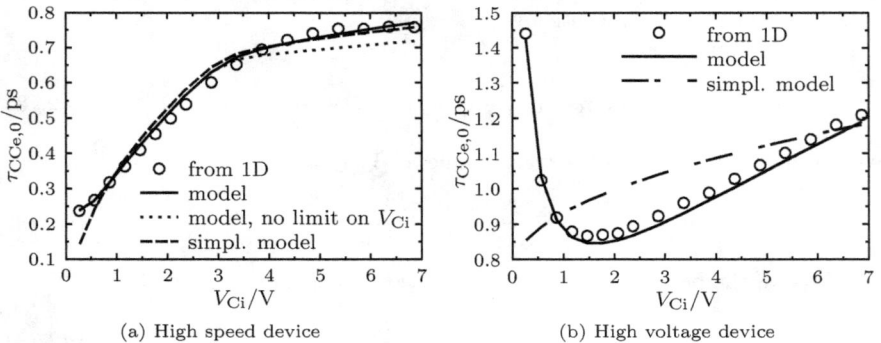

(a) High speed device

(b) High voltage device

Figure 2.63: Model for the bias dependence of $\tau_{CCe,0}$ for the high-speed and high-voltage device. In "model" the limitation of V_{Ci} to V_{PT} in the third term is applied while it is omitted for the second curve. This option is not shown for the high power device, since the shown region does not cover the punch-through case. The curve "simpl. model" completely neglects the third term and uses a modified value for $\Delta\tau_{CCe,0}$.

2.5.5 Implementation

The complete implementation is given in App. A.3 including a list of all model parameters and employed equations. In this section only the basic concepts are briefly explained.

Although (2.172) is numerically stable it has no continuous derivative. This is achieved by adding an empirical smoothing term to the equation resulting in (A.51). For the final implementation the low-injection punch-through models (2.168) and (2.178) are merged with (2.180). This is justified, because both have the same model equation and the same non-ideality factor ζ_{PT}. Furthermore, in order to avoid numerical issues the term in (2.180) correlated to ζ_{PT} is limited to 0. This is especially required for very small voltages.

Yet, no smoothing of operating regions, i.e. the partially and fully depleted collector, but a simple conditional statement is used. This does not cause numerical issues, since the equations are based on the transit times, i.e. the derivative of E_{jC}. Ensuring continuous transit times when switching from one region to another was the boundary condition when deriving the equations. Although this might be changed for a production type compact model it is sufficient for the following verification.

Finally, two possible options exists for calculating the field and the corresponding collector charge. The numerically simpler implementation is carried out *after* the calculation of I_{T}. Thus, the current dependent charge has no influence on the transfer current but will yield the desired dynamic currents. By using this approach the models for the weighted and the actual charge are not consistent anymore. The other option is the calculation inside the iteration for I_{T} and significantly increasing the computational cost. The following discussion of the model results is only focused on the simple case. A verification of the complete model including all implications on the weight factors (especially h_{f0}, cf. chapters 2.4.2.6 and 2.4.3.2) requires additional investigations which are not part of this work.

2.5.6 Application

In this section some application results are presented comprising both Si transistors from Tab. 2.2, the transistor from chapter 2.3 and a transistor from the technology in chapter 4.

2.5.6.1 Electric field

The model results for the electric field for both the HS and the HV transistor are given in Fig. 2.64[1]. For both generally a good agreement is obtained. However, larger deviations for the HV transistors exist than for the HS transistor. The punch-through voltage is much lower for the HS transistor, leading to punch-through for almost all $V_{\mathrm{B'C'}}$. Therefore, the impact of I_{PT} and τ_{PT} is less pronounced for this transistor.

---

[1]Simulations were carried out not including impact ionization.

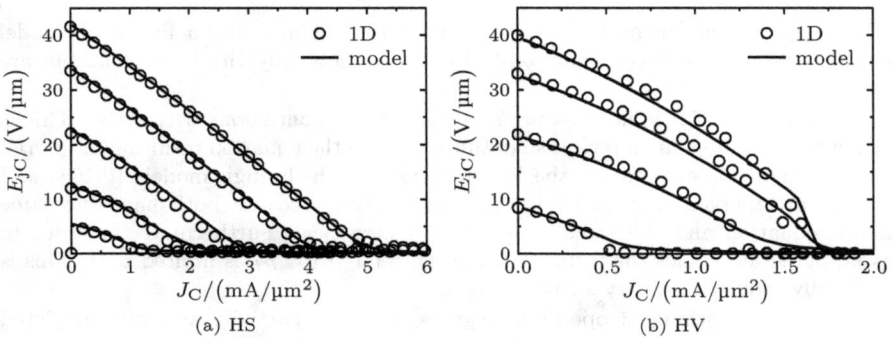

Figure 2.64: Electric field at the BC junction for (a) the HS transistor and (b) the HV transistor with $V_{B'C'}$ from $-6\,$V to $0.5\,$V.

a The resulting transit times according to (2.151) are shown in Fig. 2.65. Similar to the field model very accurate results are obtained for the HS transistor. Not using a smoothing function for linking both operating ranges causes a distinct peak in τ_{CCe} visible for small V_{Ci}. The strong increase of the transit times at I_{CK} is not modeled correctly for low V_{Ci} for the HV transistor. Only for very high voltages a partial agreement is obtained, although with an overestimation for certain voltages.

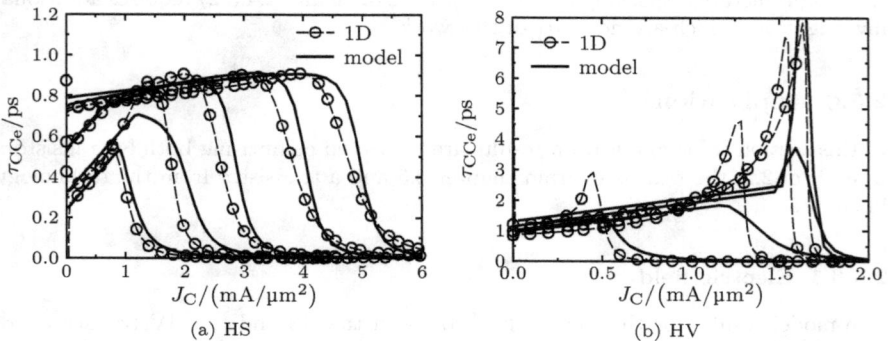

Figure 2.65: Resulting transit time for (a) the HS transistor and (b) the HV transistor with $V_{B'C'}$ from $-6\,$V to $0.5\,$V.

For both transistors the transit time caused by the reduced electric field is small compared to the base transit time. Thus, the impact on f_T is not discussed.

2.5.6.2 Small-signal parameters

As previously discussed in chapter 2.3.2.3 the imaginary part of $\underline{Y}_{21}$ is not modeled correctly by the actual HICUM/L2. The changing charge in the collector was found to be a possible cause. Shown in Fig. 2.66 is the E_{jC} model for the transistor discussed in this section. A partially accurate model can be obtained also for the SiGe HBT, although larger deviations for small voltages cannot be avoided yet. The corresponding charge leads to a dynamic current in the BC-branch and results in a strong increase of $\Im\{\underline{Y}_{21}\}$. A very accurate model for this component is obtained when using a more physics-based value for α_{IT}.

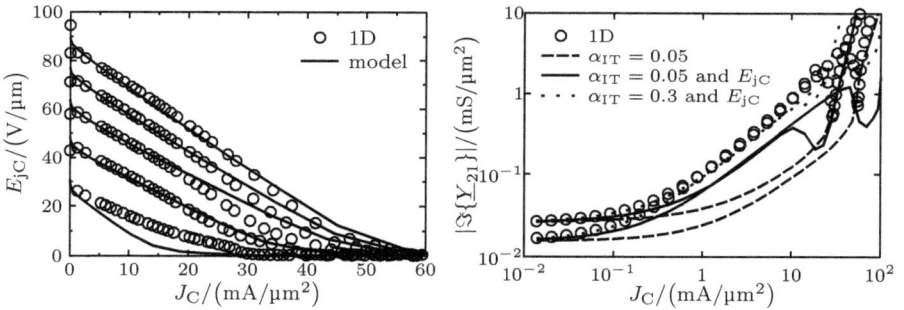

Figure 2.66: Electric field and imaginary part of $\underline{Y}_{21}$ for the transistor from chapter 2.3. $V_{C'E'}$ was swept from 0.5 V to 2.5 V.

2.5.6.3 Application to measurements

Based on the transit frequency and $\Im\{\underline{Y}_{21}\}$ the application of the field model to experimental data is visualized in Fig. 2.67. Today no dedicated extraction method exists. The transit times are the main parameters to be extracted, because the values for the low-current electric field are calculated from parameters of C_{jCi}. Those are independently obtained from conventional S-parameter measurements. The smoothing factors were chosen based on existing results for 1D simulations.

In comparison to the model extracted for HICUM/L2 without the field model a better agreement of f_T for small voltages is obtained when utilizing the field model. However, the model for high voltages suffers from utilizing the E_{jC}-model. For $\Im\{\underline{Y}_{21}\}$ only a marginal improvement in the medium current region was achieved. For complete transistor structures time constants associated with the base resistance and BC-capacitance already lead to an increasing $\Im\{\underline{Y}_{21}\}$.

These plots also depict numerical issues ($\Im\{\underline{Y}_{21}\}$ for large I_C and f_T for $V_{BC} = 0$ V) which are not addressed further in this work. For the extracted model card they are mainly related to large values of $\Delta\tau_{CCe,0}$, leading to a negative value of $E_{jC,PT}$. Additionally (although not a numerical but a model issue), $\tau_{CCe,0}$ is decreasing with

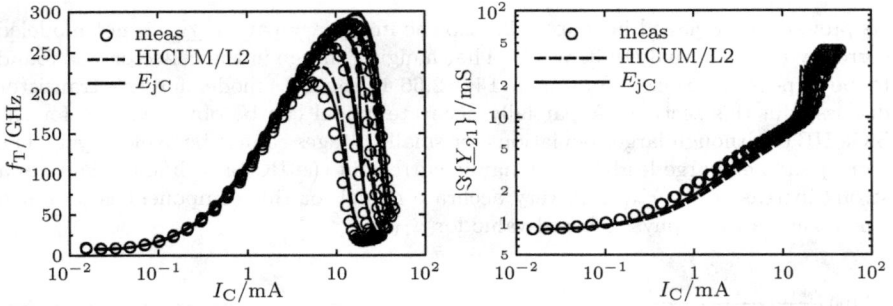

Figure 2.67: Application of the field model on experimental data. V_{BC} is swept from $-0.5\,\text{V}$ to $0.5\,\text{V}$.

V_{Ci} rather than increasing due to a disadvantageous combination of V_{lim} and V_{DCi} in (2.182). Due to this unintended behavior f_T was not modeled correctly for high voltages.

2.5.7 Summary

In this chapter the modeling approach for the electric field in the collector of a bipolar transistor based on transit times was presented. Model results compared to 1D simulations show the feasibility of such model equations. However, no useful set of parameters was extracted for actual measurements of the latest generation of high-speed SiGe HBTs.

The main issues of the model presented are

- the models for temperature dependences which are not yet developed but required for the comparison with experimental results, and

- the discussion of equations for non-constant collector doping which is the most common case especially for high-speed devices.[1].

These issues will be addressed in future work on this model.

So far no discussion of the interaction with the GICCR was presented and dedicated extraction routines are still in development. Additional effects of the heterojunction barrier, the non-constant base doping at the junction and the numerical issues that were revealed when applying the model to experimental results have also to be addressed in the future.

---

[1]Notice that the results given in 2.5.6.2 are for a transistor with non-constant collector doping.

2.6 Modeling of the substrate coupling for transistors with STI in the active region

2.6.1 Impact on small-signal characteristics

For high-speed applications substrate effects can play a significant role on the small-signal characteristics of a transistor. The n-doped collector creates in conjunction with the p-doped substrate an additional junction including diode current and depletion capacitance. Also, together with the p-doped base, a pnp-transistor with the npn-collector as base is formed. Currents caused by the SC-junction are affected by the finite permittivity of the substrate during small-signal operation due to the long path between the internal substrate-collector (SC) junction and collector contact. Discussions on the influence of the substrate coupling for SiGe-transistors were recently published in [Fis12,Ste13] but also can be found in older literature (e.g. [PRH96]).

In most practically relevant operating regions the SC-junction is biased in reverse direction. Thus, the dynamic currents caused by the depletion capacitance are the main focus of this chapter. Although the dynamic currents are mainly affecting the imaginary part of small-signal parameters, also the real part is affected due to the feedback on the substrate and collector resistance. This effect is visualized in Fig. 2.68 for a transistor with and without substrate coupling. For the simulation setup refer to 2.6.5.1. While only an increase in the imaginary part with maintained frequency dependence is obtained, substrate effects significantly affect the real part of $\underline{Y}_{22}$ and therefore directly relevant FOMs such as $f_{\max}$.

As also shown in Fig. 2.68 all substrate effects of a transistor can be represented by a simple passive structure containing only the pn-junction from substrate to collector. Thus, a correct model for the substrate-collector junction is sufficient for characterizing the substrate coupling. However, at large collector currents a lateral voltage drop across the highly doped collector from the contact towards the internal transistor exists, leading to a spatial dependence of the substrate capacitance. In this case the assumption that the separate substrate structures provides the correct results in comparison to the real transistor is no longer valid. This effect is not discussed here but will be shown in more detail in [Zim15]. However, for highly doped buried layers this voltage drop is quite small, allowing to neglect this effect for most transistors.

However, when utilizing the resistance values extracted from the frequency dependences at reverse bias also at forward bias, no accurate curves for the DC currents are obtained. The reasons behind this effects are also discussed briefly in this chapter. The model developed is required to find unique parameters for both DC and AC operation.

2.6.2 Structure and equivalent circuit

The evaluated structures are sketched in Fig. 2.69. There, two distinct cases are discussed. In the first case (a) an overlapping collector is assumed, i.e. the junction

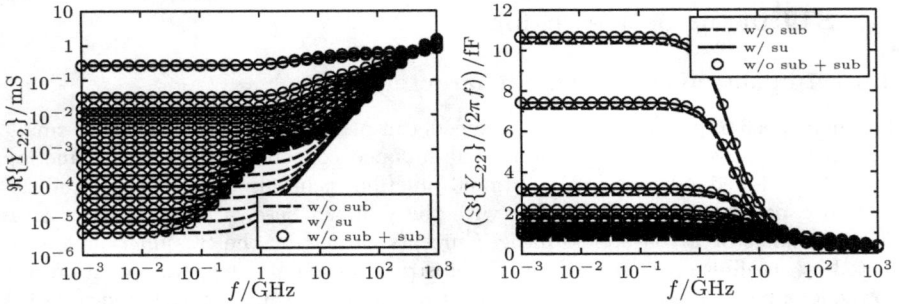

Figure 2.68: Real and imaginary part of $\underline{Y}_{22}$ for a transistor with and without substrate coupling. The curves correspond to operating points from $V_{\mathrm{BE}} = 0.7\,\mathrm{V}$ to $1.0\,\mathrm{V}$. Here, "w/o sub" corresponds to 2D simulations of the transistor only, while "w/ sub" refers to results from the complete transistor including substrate coupling. "w/o sub + sub" is related to curves calculated from $\underline{Y}_{22}$ of the "w/o sub" transistor and the small-signal admittance of the corresponding stand-alone passive substrate structure.

extends below the STI-oxide. When merging most distributed effects into lumped elements the junction capacitance can be separated into an area related component C_{jSa} with a series resistance $R_{\mathrm{su,a}}$ and a perimeter related component C_{jSp} with a series resistance $R_{\mathrm{su,p}}$. The STI-oxide is causing a capacitance C_{STI} which is in parallel to the current path related to the depletion capacitance. Additionally, a contact resistance R_{cont} can be inserted.

In the second case (b) the STI-oxide is not overlapping the collector. In the corresponding equivalent circuit the perimeter related path disappears. For simplicity the capacitances parallel to the series resistances, which are caused by the permittivity of the substrate, are omitted in the figure. This structure corresponds to a deep-trench isolation (DTI).

In HICUM/L2, substrate coupling is modeled by an RC-network in series to the depletion capacitance. The corresponding circuit is given in Fig. 2.70(a). Here, "Ci" is the internal collector node and "S" the substrate contact. The equivalent circuit proposed in this work is presented in Fig. 2.70(b). A capacitance in parallel to the existing path as well as the contact resistance R_{cont} are added. Since this resistance adds an additional node to the compact model a short discussion of the component is given later.

The proposed EC which can also be found in literature (e.g. in more complicated form in [PRH96]) corresponds to the structure in Fig. 2.69(b) but can be applied to the structure from Fig. 2.69(a), too. In the latter case C_{SCp} is the sum of C_{STI} and C_{jSp}, while R_{cont} is the effective value from merging the actual R_{cont} with $R_{\mathrm{su,p}}$.

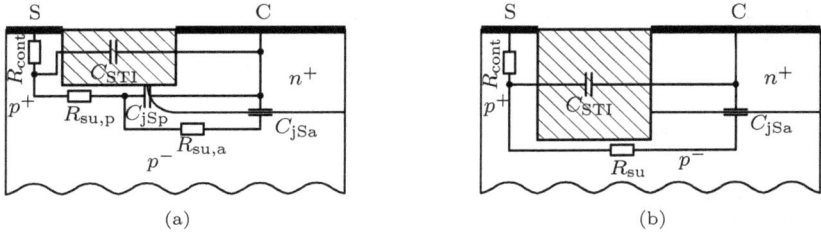

Figure 2.69: Cross section of the assumed structure for (a) a collector overlapping the STI-oxide and (b) a collector not overlapping the STI-oxide.

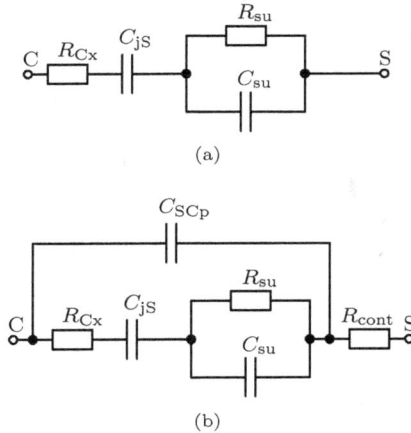

Figure 2.70: (a) Existing equivalent circuit for the substrate coupling in HICUM/L2. (b) Proposed equivalent circuit.

However, due to the effects for large widths of the STI described in the next section C_{SCp} can also consist of C_{STI} and C_{jSp}, while C_{jS} is only defined by C_{jSa}. Since both applications are practically relevant the model equation for C_{SCp} also contains a bias dependence of a depletion capacitance. In order to verify the employed EC the more complicated structure from (a) is used for further discussions.

For application of the equivalent circuit in a bipolar compact model, the external collector resistance also needs to be included. However, the actually simulated test-structure is not capable of shown in the correct influence of R_{Cx}. It will therefore not be discussed here further.

2.6.3 Geometry scaling

For evaluating the substrate coupling 2D numerical device simulations of the SC-junction including a shallow trench isolation according to Fig. 2.69(a) are performed. In the subsequent pictures containing stream lines the STI is indicated by thick lines. The dimensions roughly align with the technology discussed in chapter 4.1.

For low frequencies the complete substrate collector capacitance C_{SC} consists of the depletion capacitance C_{jS} and a capacitance C_{STI} caused by the shallow trench oxide. The depletion capacitance is furthermore separated into an area (C_{jSa}, below the collector contact) and a perimeter (C_{jSp}, below the STI oxide) component.

In this chapter, length specific quantities are labeled by Q' and area specific components by $\overline{Q}$. Since only 2D simulations were performed, all measured quantities are length specific.

Separation of area and perimeter component

The scaling equation

$$C'_{\text{SC}} = f(b_{\text{C}}) = C'_{\text{SCp}} + \overline{C}_{\text{jSa}} b_{\text{C}} \tag{2.188}$$

is applied. As shown in Fig. 2.71(a) the approach of a linear scaling with the buried layer width b_{C} is justified. From this scaling law C'_{SCp} as a function of b_{STI} can be extracted based on different widths of the STI-oxide. For not too narrow (and unrealistic) widths of the oxide the same value for $\overline{C}_{\text{jSa}}$ is obtained independent of b_{STI}. C'_{SCp} is extracted from the y-axis intercept point of (2.188). The results for the latter for different b_{STI} are shown in Fig. 2.71(b). Since $\overline{C}_{\text{jSa}}$ is independent of b_{STI}, C'_{SCp} can be directly extracted from structures with different b_{STI} and the same b_{C}.

The perimeter component splits into two parts

$$C'_{\text{SCp}} = C'_{\text{STI}} + C'_{\text{jSp}}. \tag{2.189}$$

One part results from the buried layer doping which has a larger depth than the STI oxide for the simulated structure. Thus, it is creating a junction underneath the STI oxide (C'_{jSp}). The other part is caused by the capacitance of the oxide itself and strongly depends on b_{STI} (C'_{STI}).

For narrow oxides the perimeter related capacitance increases with decreasing width, while for increasing widths a constant value is reached. The reason behind

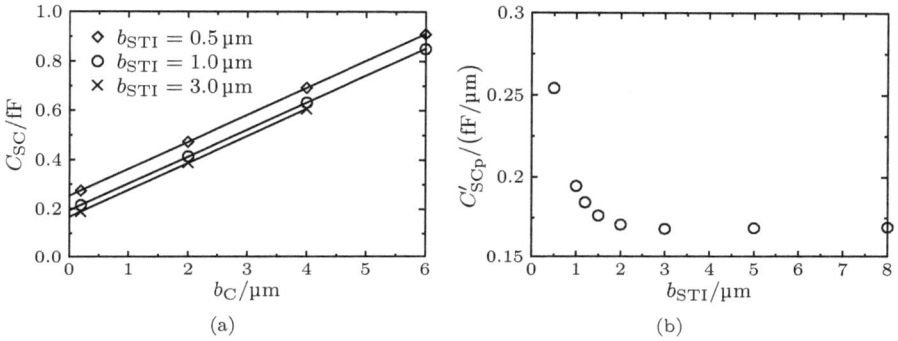

(a)

(b)

Figure 2.71: (a) Scaling of C'_{SC} versus b_C at $V_{SC} = 0\,V$ for different widths of the STI-oxide. (b) Values of C'_{SCp} obtained from scaling and application of (2.188) versus b_{STI}.

this behavior is visualized in Fig. 2.72. For narrow oxides the STI-oxide can be seen as a plate capacitor. In this case its capacitance is defined by b_{STI}. However, for wider oxides large parts of the region below the oxide become neutral resulting in a virtual contact. The field lines through the oxide are terminating near the end of the depletion zone. Thus, in this case the capacitance is almost independent of b_{STI}. Based on conformal mapping, especially Schwarz-Christoffel-mapping, the capacitance of the oxide region can be calculated and, thus, also its limit for $b_{STI} \to \infty$. A detailed application is given in [Zim15].

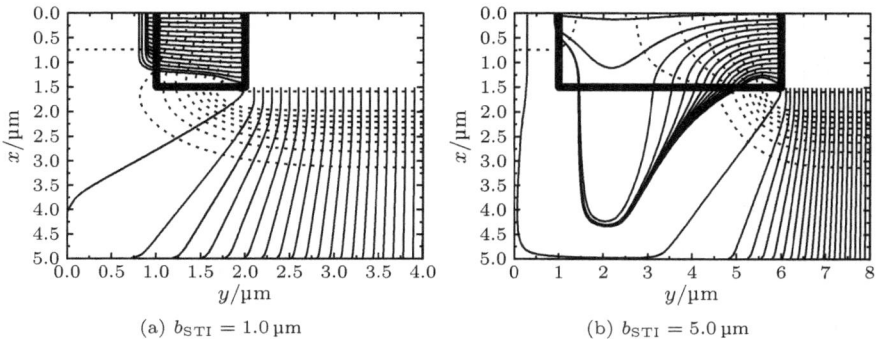

(a) $b_{STI} = 1.0\,\mu m$

(b) $b_{STI} = 5.0\,\mu m$

Figure 2.72: DC field lines (solid) at $V_{SC} = 0\,V$ for different widths of the STI oxide. Equipotential lines are included to visualize the SCR (broken lines).

Split of perimeter component

Because C'_{STI} is behaving as a plate capacitor for small b_{STI} a separation of C'_{SCp} into its two components can be performed as described in the following. C'_{STI} is caluclated as

$$C'_{\text{STI}} = \frac{k_{\text{C}}}{b_{\text{STI}}}, \tag{2.190}$$

with $k_{\text{C}} = w_{\text{STI}}\varepsilon$ and the depth of the oxide w_{STI}. It satisfactorily models the capacitance for narrow oxides. Using the extracted values of C'_{SCp} for at least two narrow oxides and a linear optimization versus $1/b_{\text{STI}}$ allows to extract both k_{C} and the perimeter component of the depletion capacitance C'_{jSp} from the slope and intercept of (2.190).

For C'_{STI}, the following simple model is proposed here. An effective distance

$$b_{\text{STI,eff}} = b_{\text{STI,con}} - r_{\text{STI}} \frac{x_{\text{STI}} + \sqrt{x_{\text{STI}} + 0.01}}{2}, \tag{2.191}$$

is calculated with

$$x_{\text{STI}} = \frac{b_{\text{STI,con}} - b_{\text{STI}}}{r_{\text{STI}}}, \tag{2.192}$$

and $b_{\text{STI,con}} = k_{\text{C}}/C_{\text{STI,con}}$. Here, $C_{\text{STI,con}}$ is the constant capacitance of the oxide for large b_{STI}, while r_{STI} is an empirical smoothing parameter. $b_{\text{STI,eff}}$ is inserted into (2.190) instead of the actual b_{STI}. An application of this model with $r_{\text{STI}} = 3$ is given in Fig. 2.73 showing a very good agreement in the transition region of C_{STI}.

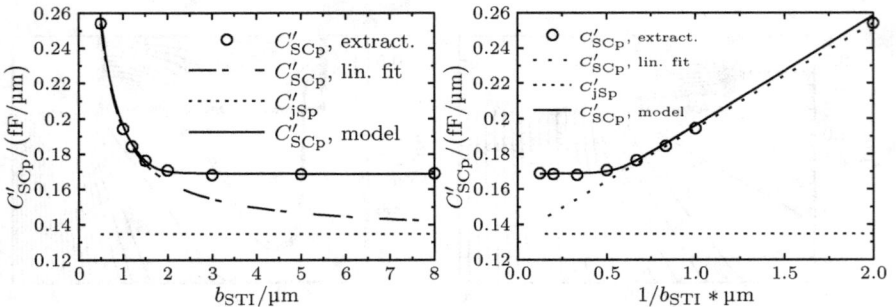

Figure 2.73: Modeling the perimeter component C'_{SCp} with C'_{jSp} and C'_{STI} employing (2.190) and (2.191). Note, "lin. fit" represents the linear optimization with respect to $1/b_{\text{STI}}$. The right picture emphasizes the assumption of a plate capacitor.

Series resistances

For the equivalent circuit in Fig. 2.70(b), the real part of the substrate impedance ($\underline{Z}_{SC} = 1/\underline{Y}_{SC}$) simplifies for low frequencies to

$$\lim_{f \to 0} \Re\{\underline{Z}_{SC}\} = R_{cont} + R_{su} \left(\frac{C_{jS}}{C_{jS} + C_{SCp}} \right)^2. \tag{2.193}$$

Note again that the external collector resistances is not discussed here in this chapter.

The series resistance below the STI-oxide can be extracted for very wide oxides and low frequencies. In this region all distributed effects below the oxide can be neglected, allowing to extract the resistance from (2.193) versus b_{STI}. The scaling is visualized in Fig. 2.74 showing the anticipated linear slope.

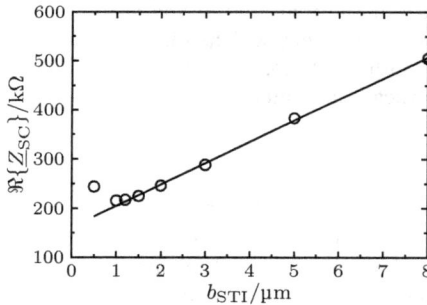

Figure 2.74: Scaling of the resistance from (2.193) versus b_{STI} for $b_C = 2\,\mu m$.

For very narrow oxides the resistance is again increasing with reducing b_{STI}. The reason for this non-physical behavior is originated from the simulation setup using a spatially limited region and visualized in Fig. 2.75. Due to the finite simulation region in Fig. 2.75(a) the current path becomes narrow as soon as the SCR overlaps the STI oxide. This simulation setup can be interpreted as part of a transistor array where one substrate contact is shared by two transistors on either side. The opposite case of an "infinite" simulation region next to the contact in Fig. 2.75(b) does not show this effect because the location of the flow lines is not restricted. Due to these non-ideal and complicated effects no scaling law for R_{su} is presented here.

Results

The equivalent circuit from Fig. 2.70(b) is applied to the small-signal resuls of the simulated structures. The bias dependences of $\overline{C}_{jSa}$ and C'_{jSp} are extracted by applying (2.188) for several V_{SC}. Also, bias dependent values were extracted for C'_{STI} and R_{su}. These result from the shift of the SCR edges with changing voltages and are further neglected.

Results are given for a structure with $b_C = 2\,\mu m$ and $b_{STI} = 1\,\mu m$. For this relatively small STI-oxide both components of the depletion capacitance are merged

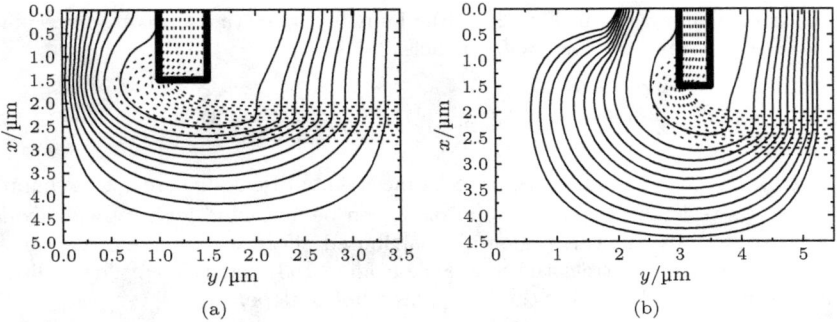

Figure 2.75: Stream lines of the real part of the current density at $V_{SC} = 0\,\mathrm{V}$ and $f = 1\,\mathrm{MHz}$ for different widths of the simulated structure next to the substrate contact. Note, the stream lines for the imaginary part are almost identical and, thus, not shown.

into C_{jS} while $C_{SCp} = C_{STI}$ is used. Moreover, the oxide creates an almost ideal plate capacitor distinctly separating C_{STI} from C_{jSp}.

As highlighted by the results displayed in Fig. 2.76 the proposed EC accurately models the small-signal characteristics of the substrate region. Using only the parallel capacitance without R_{cont} provides good results below $100\,\mathrm{GHz}$. For higher frequencies the additional R_{cont} has a significant effect on the real part. Although the distributed character of the substrate effects yields non-ideal curves (i.e. the increase of $\Re\{\underline{Y}_S\}$ is not following f^2 for high frequencies), utilizing lumped elements provides acceptable results in the complete investigated frequency range.

2.6.4 DC modeling

In DC operation R_{su} (and R_{cont} for the extended EC) acts as a feedback resistance for the substrate diode and transistor. At forward operation significant conductivity modulation occurs due to the injected electrons and the low doping of the substrate. The following derivation is based on 1D simplifications but can also be applied to 2D/3D.

A simple R_{su} model can be obtained based on the injected electrons similar to the model for the base resistance [SC10]. The movement of the borders of the SCR are not taken into account in order to keep equations simple. Based on the hole transport equation the resistance is given by

$$R_{su} = \frac{1}{\int_{x_s}^{w_s} A_S q \mu_p p} \, \mathrm{d}x, \tag{2.194}$$

with the hole mobility μ_p and the relevant area A_S. x_s is the beginning of the

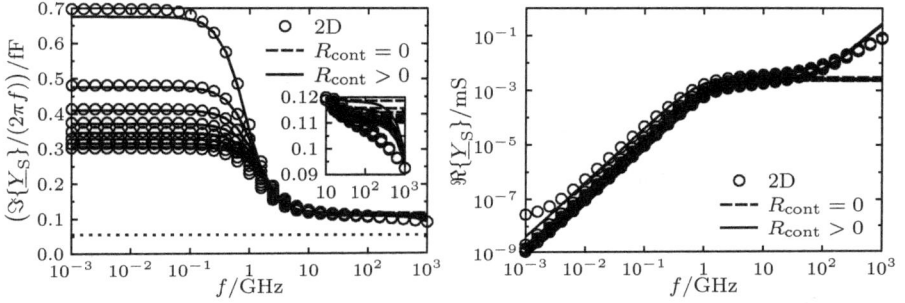

Figure 2.76: Application of the proposed equivalent circuit on the small-signal results for the capacitance structure with $b_C = 2\,\mu\mathrm{m}$ and $b_{STI} = 1\,\mu\mathrm{m}$. V_{SC} was swept from $-1.0\,\mathrm{V}$ to $0.4\,\mathrm{V}$. In the left plot C_{STI} is added as dashed line.

neutral region and w_s the end of the substrate, i.e. the contact location. Based on the Poisson equation for neutral regions above equation is rewritten into a normalized form

$$R_{su} = R_{su0} \frac{Q_{su0}}{Q_{su0} + Q_{n,su}}, \tag{2.195}$$

where R_{su0} and

$$Q_{su0} = \int_{x_s}^{w_s} A_S q p \, dx, \tag{2.196}$$

are the resistance and hole charge in the substrate path at low injection, respectively. $Q_{n,su}$ is the injected electron charge. The latter is calculated from the injected electron density n_s by

$$Q_{n,su} = \int_{x_s}^{w_s} A_S q n_s \, dx. \tag{2.197}$$

The bias dependence of n_s is given by

$$n_s = \frac{N_S}{2} \left[\sqrt{1 + 4\frac{n_i^2}{N_S^2} \exp\left(\frac{V_{S'C'}}{V_T}\right)} - 1 \right], \tag{2.198}$$

with the doping concentration in the substrate N_S. Thus, the model equation for $Q_{n,su}$ reads

$$Q_{n,su} = \frac{Q_{su0}}{2} \left[\sqrt{1 + 4\frac{I_{SCs}}{I_{Ks}} \exp\left(\frac{V_{S'C'}}{V_T}\right)} - 1 \right]. \tag{2.199}$$

Here, the knee current I_{Ks} is a model parameter. Note that $I_{SCs} \exp\left(V_{S'C'}/V_T\right)$ is used rather than $I_{SCs} \exp\left(V_{S'C'}/(m_{SC}V_T)\right)$, because the non-idealities summarized in m_{SC} are not affecting the injection.

Inserting (2.199) into (2.195) leads to the final model equation

$$R_{su} = R_{su0} \frac{2}{1 + \sqrt{1 + 4\frac{I_{SCs}}{I_{Ks}} \exp\left(\frac{V_{S'C'}}{V_T}\right)}}, \quad (2.200)$$

with I_{Ks} as only remaining model parameter.

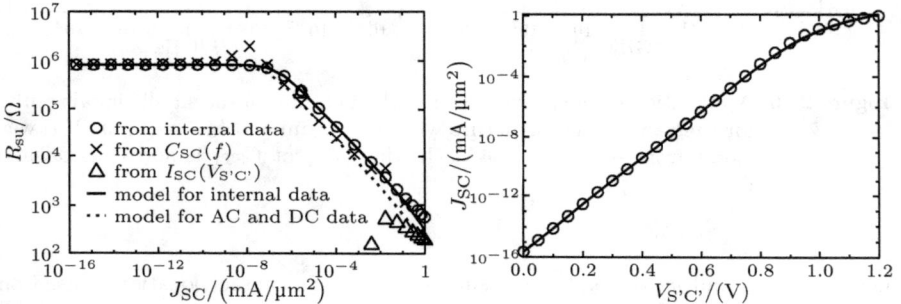

Figure 2.77: Modeling of R_{su} with (2.200). The parameter I_{Ks} was extracted from internal data as well as from small-signal and DC data. The model for I_{SC} utilized the R_{su} value extracted from small-signal and DC data.

The extraction of R_{su} can be performed based on either the small-signal $C_{jS}(f)$ curves or the $I_{SC}(V_{SC})$ curve. While the former is well suited for low and reverse bias, very low frequencies are required for large voltages. In this range, though, the DC current extraction can be performed by using the standard diode equation and calculating the voltage drop. All methods in comparison to the actual value based on (2.194) are shown in Fig. 2.77, providing also the application of (2.200) to the differently extracted values. The equation is capable of providing good models in any case. The difference in the extracted values are caused by the neglected hole current which is not affected by high injection effects. Shown in this picture is also the final application to model the DC current yielding accurate results.

For the extended EC R_{cont} can still include a part of the lowly doped substrate. Hence, its value at $V_{SC} = 0\,\text{V}$ may easily exceed the bias dependent R_{su} at high forward bias. Therefore and for reasons of simplicity R_{cont} is modeled in the same way as R_{su}, i.e. using (2.200).

2.6.5 Application to transistors

The application to small-signal characteristics of transistors is given for 2D numerical device simulations and actual measurements. The former are based on the structures from 2.6.3 including a realistic vertical and lateral profile for a transistor.

2.6.5.1 Numerical simulations

The substrate capacitance is obtained from measured Y-parameters by

$$\underline{Y}_S = \underline{Y}_{22} + \underline{Y}_{12} \tag{2.201}$$

and

$$C_{SC} = \frac{\Im\{\underline{Y}_S\}}{2\pi f}. \tag{2.202}$$

Note that (2.201) cannot be applied to Y-parameters during forward operation at high bias. There, the output conductance g_o dominates the real part of $\underline{Y}_{22}$ and also strongly impacts the imaginary part. Using $\underline{Y}_{21}$ rather than $\underline{Y}_{12}$ also leads to wrong results due to the current dependence of C_{jC}. Thus, reasonable values for $\underline{Y}_S$ as a function of V_{SC} can only be obtained in operation points outside the active region, e.g., by utilizing standard S-parameter measurements for C_{BC} with $V_{BE} = 0\,\mathrm{V}$.

Extraction results of the substrate admittance from transistor characteristics are given in Fig. 2.78. As discussed before a sweep of V_{BC} with $V_{BE} = 0\,\mathrm{V}$ is preferred over a forward gummel sweep. For low frequencies $\underline{Y}_S$ extracted from (2.201) yields the same results as the passive substrate structure. However, at high frequencies deviations occur due to the npn transistor. Thus, $\underline{Y}_S$ obtained from (2.201) cannot be used in the complete frequency range. Since the effect is related to the actual transistor a previously extracted model without substrate coupling can be used to perform a correction.

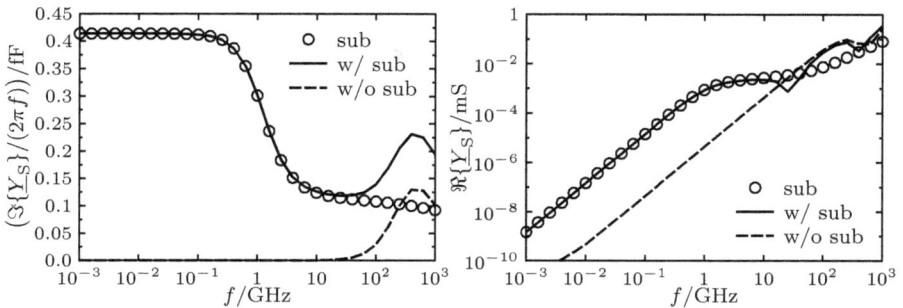

Figure 2.78: Extracted $\underline{Y}_s$ from a transistor with (2.201). Values are shown for $V_{BE} = V_{BC} = 0\,\mathrm{V}$. In both plots "w/o sub" and "w/ sub" are similarly defined as in Fig. 2.68. "sub" presents results for the substrate structure.

In this chapter's introduction the curves for $\underline{Y}_{22}$ for the transistor with substrate effects were already shown. As discussed there briefly, an accurate model for the separate substrate structure yields correct results. The substrate structure dimensions in 2.6.3 were chosen according to the transistor from Fig. 2.68. Since in Fig. 2.76 the accurate model of the substrate admittance is shown, the Y-parameter is not repeated here.

2.6.5.2 Measurements

Low frequency measurements are required for the extraction of the accurate substrate coupling model. Since these were not available at the time of this work results are only shown for a single transistor in CBEBC configuration. Scaling results are omitted, because the extraction of all relevant parameters is too error prone to allow a verification of the equations.

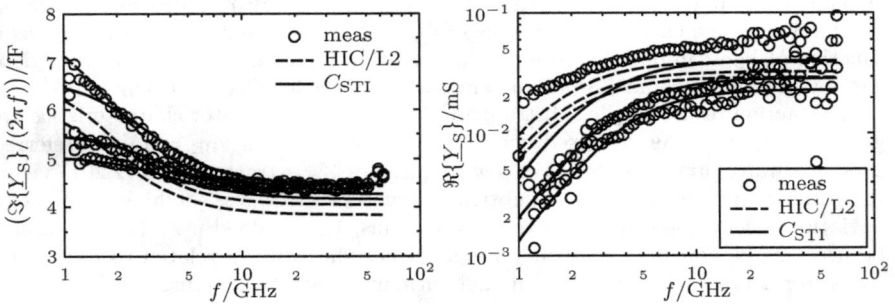

Figure 2.79: Extracted $\underline{Y}_s$ with (2.201) from actual transistor measurements. The plots are given for $V_{SC} = [-1.0, -0.5, 0]$ V.

The comparison of the model for $\underline{Y}_S$ is given in Fig. 2.79 showing a significant improvement of the imaginary part especially for high frequencies. Also, the bias dependence of the real part is modeled more accurately due to the extended EC. The impact on the modeling of $\underline{Y}_{22}$ is shown in Fig. 2.80. For low current densities, i.e. where the real part of $\underline{Y}_{22}$ is not dominated by g_o, an accurate model is obtained. Note that due to the limited available frequency range R_{cont} was not extracted.

2.6.6 Summary

In this section the substrate modeling with HICUM/L2 was discussed. Based on an existing technology with a STI around the active region a physics-based yet simple extension of the EC was presented, providing an accurate model for both 2D simulations and actual measurements. In its simpler form, i.e. without an additional contact resistance, no additional node is required. Only a capacitor is placed in parallel to the existing substrate coupling path. This capacitance is mainly caused by the STI oxide but will also be required for processes with deep-trench isolation (DTI). For an STI process this capacitance can contain portions of the peripheral depletion capacitance so a bias dependence was implemented. The actual mapping of the extracted physical components to the components in the EC strongly depends on the actual geometry of the STI and the collector doping.

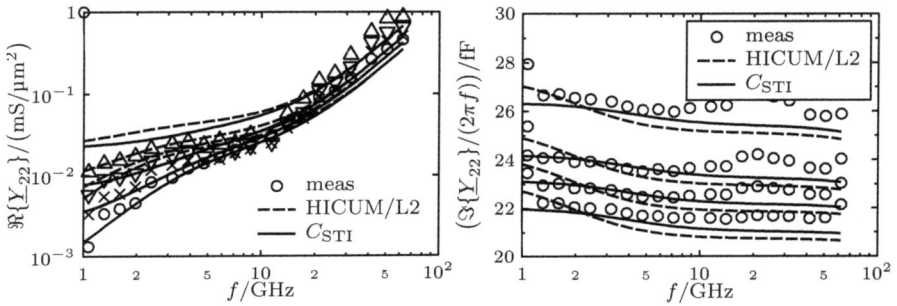

Figure 2.80: Modeling of $\underline{Y}_{22}$ for actual measurements with the existing HICUM/L2 EC and the proposed EC for $J_C = [0.1, 0.5, 1, 2]\,\text{mA}/\mu\text{m}^2$.

Some simple scaling rules for the substrate capacitance with the extended EC were presented. Employing those allows to accurately extract the different components of the capacitance as presented by results based on 2D simulations. However, due to the limited available frequency range especially towards low frequencies they were not verified by experimental results.

3 | Compact model parameter extraction methodology

3.1 Extraction flow

The general extraction flow for a scalable bipolar technology is given in Fig. 3.1.
Before starting with the actual parameter extraction, all process relevant geometry
parameters should either be extracted from cross-sections taken from, e.g., TEM
measurements or transistor layouts. A schematic cross-section of a bipolar transistor with a selection of relevant dimensions is given in Fig. 3.2. Based on those
parameters, elaborate scaling equations allow to accurately calculate specific model
parameter. In this work, TRADICA ([SZ06]) is employed for the calculation of
several compact model parameters, including series resistances and transit times.

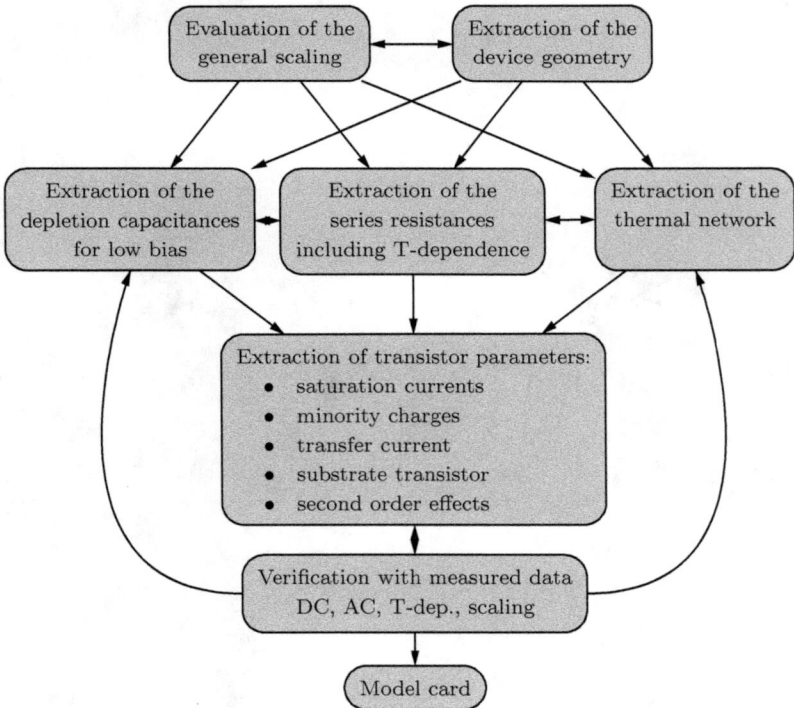

Figure 3.1: General steps of the parameter extraction procedure for a geometry scalable process. Here, the focus is on the characterization of devices. Thus,
additional mandatory steps for production type model cards (e.g. stati-
cal distribution, corner models) are not included in the chart.

Beside obtaining the dimensions of the technology as described above, initial

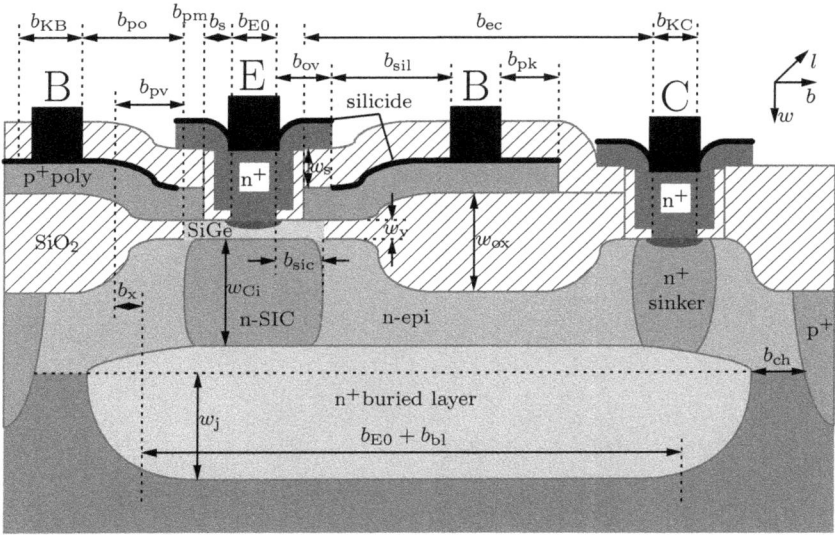

Figure 3.2: Schematic cross-section for a biploar transistor with relevant lateral dimensions. Reprint from [SZ06,Kra15].

scaling investigations on at least the collector current are performed. This is required to ensure the usage of correct physical dimensions for the emitter window, which is the basis for almost all scaling equations.

Extraction of the (low bias) capacitances, series resistances and the thermal network are generally performed before extracting parameters for currents and transit times of the transistor. Series resistances (at least R_B and R_{Cx}) are often extracted from dedicated test structures rather than from actual transistors, although test structures for R_E are often not possible in advanced SiGe technologies ([KS14]). Furthermore, the extraction of the capacitances and the thermal resistance does not require the previous extraction of other transistor parameters. However, parameters required for e.g. for the extraction of the bias dependent R_B are previously obtained from capacitances. Thus, the steps require interaction to some extent.

However, almost all remaining transistor parameters, e.g., parameters for the transfer current and the minority charge, require accurately known values for the capacitances and resistances and are thus extracted in a following step. After extraction of all relevant parameters, verification of the compact model with respect to the measured data is performed. The minimum of verification requires to compare DC characteristics, i.e. terminal voltage-current curves, and small-signal S- or Y-parameters as function of frequency and operating point, including derived quantities such as the depletion capacitances, f_T and f_{max}. The verification is performed over a large bias and temperature range, though it can be limited by the available measurement equipment.

Finally, most extraction steps including fine-tuning of extracted parameters are repeated until the accuracy of the modeled characteristics with respect to the measured reference has reached a certain - although always user defined - level.

3.2 Joint extraction of the parasitic emitter and the thermal resistance

In this work, a novel extraction methodology for the emitter resistance R_E was developed. As a consequence of the employed model equations, the thermal resistance can also be extracted by this method.

Generally, extraction methods for R_E exist based on DC characteristics as well as small-signal parameters. The latter always employ a more or less simplified equivalent circuit of the bipolar transistor. Examples can be found in the classical Gobert method ([GTB97]) or in more complicated methods as described in [HS09] and [RAH11]. Though quite accurate results can be obtained by employing those methods, the application is limited by two facts. First, in contrast to DC methods, not only the bias region but also the frequency must be selected very carefully. While for too low frequencies dynamic self-heating affects results at high currents, too large frequencies might result in influences from RC-time constants, which are not taken into account by the specific equivalent circuit. The second restricting factor is the availability of equipment for the required frequency band or possibly limited measurement accuracy.

Although methods based only on the terminal DC current and voltage characteristics do not face these issues, generally simplified (semi-)physical equations are applied to find a dependence of transistor characteristics on R_E. Often employed methods include [NT84] and the open-collector method [GR98]. As a limiting fact in most of these methods, in [KS14] the strong influence of thermal effects on the extracted values for DC methods is shown. However, as shown for example in [SRVC14], the operating current densities of the transistors are gradually increasing from one generation to the next, increasing the impact of thermal effects. Thus, methods where those effects are not a limiting factor are required for an accurate extraction of R_E. One method without this limitation is [TSW+97]. There, the iterative calculation of R_E and R_{th} is based on the correction of I_C and I_B by removing their increase caused by self-heating.

Based on the ideas of the latter publication, an extraction method with an even simpler and thus more stable iteration scheme was developed and is explained in the next sections. In this work, the method is applied to characterize the HBTs from the selected BiCMOS technology (4.2.1.3). Results for the application on a large set of SiGe technologies are presented in [KS14]. The application to InP HBT technologies is given in [NKS14].

3.2.1 Theory

The basic idea behind the method is using the terminal current-voltage characteristics as temperature sensor and to couple the extracted temperatures with a self-heating model in order to achieve consistent results. Since thermal effects generally occur at the same current densities where also high injection and high current effects begin, using the collector current for extracting R_E is avoided here. Although very elaborate equations exists for describing both the bias and temperature dependence (see 2.4 and the references of this chapter), applying this model will be too error prone for extracting the device temperature, since the value of the thermal resistance is generally unknown.

In contrast, the model for the base current in the forward operating region applied in compact models is quite simple, reading

$$I_{BE}(V_{B'E'}, T) = I_{BEs}(T) \exp\left(\frac{V_{B'E'}}{m_{BE}V_T(T)}\right). \tag{3.1}$$

In this equation, only the saturation current and the thermal voltage are considered as temperature dependent, while the non-ideality factor m_{BE} is assumed to be constant. Thus, providing the previous extraction of $I_{BEs}(T)$ allows using $I_{BE}(V_{B'E'})$ as a temperature sensor.

As shown in, e.g., [SC10], m_{BE} is the sum of several physical effects. It can easily be shown during extraction of I_B from device simulations that m_{BE} *is* dependent on the temperature. Only very accurate measurements on a thermal chuck (deviation of less than $1\,K$ for temperatures up to $100\,°C$ are required) would allow to extract the correct curve of $m_{BE}(T)$, which is not given by the equipment available for this work. However, utilizing an elaborate model for $m_{BE}(T)$ would improve the accuracy of this method.

The terminal I_B can accurately be described by (3.1) only in a limited range of forward operation. For low $V_{B'E'}$, recombination in the BE-SCR or tunneling currents lead to an additional non-ideal base current component, while at high $V_{B'E'}$ this value is different from the terminal V_{BE} due to the voltage drop across the series resistances. After the onset of high current effects, the strongly increased charge in the base leads to a large current component caused by neutral base recombination (NBR). The latter will also affect $I_B(V_{B'C'})$ at forward operation, which is not included in (3.1). At small $V_{C'E'}$, saturation effects start affecting I_B while at high $V_{C'E'}$, avalanche breakdown (II) results in an additional component for the base current.

The internal base emitter voltage is given by the voltage drops across the series resistances

$$V_{B'E'} = V_{BE} - I_E\left(R_E + \frac{1}{1+\beta_f}R_B\right), \tag{3.2}$$

where in the following a fairly large current gain is assumed, allowing to neglect the influence of R_B[1]. Thus, assuming a known R_E including its temperature dependence allows to use (3.1) based on the terminal $I_B(V_{BE})$ in the bias ranges as explained above.

For this method, the simple self-heating model

$$\Delta T = R_{th}I_C V_{C'E'} = R_{th}I_C\left(V_{CE} - I_C R_{Cx} - I_E R_E\right) \tag{3.3}$$

is employed. When neglecting in this equation the temperature dependence of R_{th},

---

[1] At least fo SiGe-HBTs. For technologies with a smaller current gain, accurate values for the voltage drop across the base resistance are required.

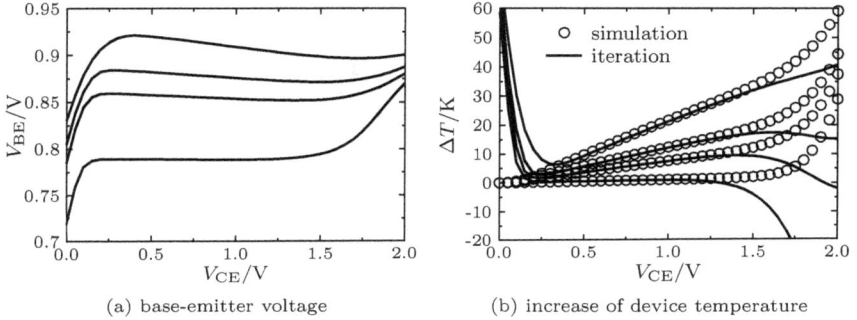

(a) base-emitter voltage

(b) increase of device temperature

Figure 3.3: V_{BE} from a compact model simulation of output curves with forced $J_{\mathrm{B}} = [1, 10, 20, 50]\,\mu\mathrm{A}/\mu\mathrm{m}^2$. Shown on the right side is the simulated temperature increase for the same bias as in (a) as well as the calculated values using (3.1) with (3.2) ("iteration").

R_{Cx} and R_{E} and assuming again a large current gain (and thus $I_{\mathrm{C}} \approx I_{\mathrm{E}}$), the temperature increase is linear with V_{CE}, for a constant I_{C}. This linear increase is visualized in Fig. 3.3(b) for output curves with forced I_{B}. Even for large self-heating where additionally the series resistances and R_{th} cannot be assumed as constant anymore, an almost ideal linear increase with V_{CE} exists. In this region, the calculation of ΔT using (3.1) with (3.2) still provides a good agreement with the actual values. Outside of this region for low V_{CE}, the calculated temperature is wrong, since the reverse base current, which leads to lower V_{BE}, is compensated by a large temperature increase in the calculation. For high V_{CE} the opposite happens due to the increase of V_{BE} caused by breakdown.

For a linear increase of ΔT, an extrapolation of the latter with respect to V_{CE} is performed towards $\Delta T = 0\,\mathrm{K}$. From (3.3) follows a characteristic voltage for $\Delta T = 0\,\mathrm{K}$ given by

$$V_{\mathrm{CE},0} = I_{\mathrm{C}} R_{\mathrm{Cx}} + I_{\mathrm{E}} R_{\mathrm{E}}. \tag{3.4}$$

Assuming a known value for R_{Cx} from previous extraction steps (e.g. based on sheet resistance structures), $V_{\mathrm{CE},0}$ is directly related to R_{E}. The extraction method for R_{E} is based on an iterative solution using (3.1) as temperature sensor, which allows the extraction of R_{E} from (3.4). The extrapolation is visualized in Fig. 3.4. Here, the extrapolation is performed using the actual values of ΔT from the simulation. The increase of $V_{\mathrm{CE},0}$ with I_{B} and thus I_{C} and I_{E} can be seen in the figure.

3.2.2 Measurement setup

The extraction method requires a V_{CE} sweep during which I_{C} is kept constant. Different methods can be applied to achieve this. A simple solution in terms of the employed measurement setup is to use output curves with forced I_{B}, which

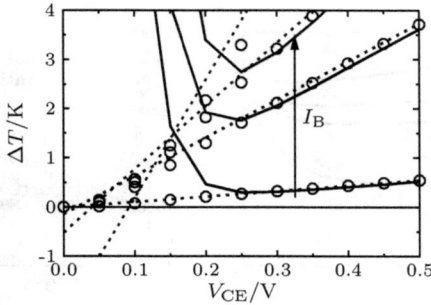

Figure 3.4: Extrapolation toward $V_{CE,0}$ from ΔT in the linear region and the same
bias as in Fig. 3.3.

was done in this work. The advantage is that standard equipment can be used
and no special test structures are required. Although the method is based on pure
DC data, a standard GSG-pad configuration with base and collector connected to
signal is sufficient. A DC deembedding (4.2.1.3) should be performed in advance to
avoid influences of the interconnections. For SiGe-HBTs with a small temperature
coefficient of the current gain, a sufficiently constant I_C is obtained.

As another option, forced V_{BE} can be applied. The advantage of this setup is
that the sweep does not have to be adjusted for each device, while the sweep of
forced I_B depends on the emitter size of the specific device. However, for this setup
I_C will increase with V_{CE} due to self-heating, leading to a non-linear increase of ΔT.

A fairly large range of ideally constant I_C is obtained from measurements with
forced I_E. However, this setup has the drawback of requiring dedicated test structures or at least a DC configuration with an accessible emitter contact. Furthermore,
the extraction range is not only limited by a linear I_C but also by (3.1) as the physical description of the base current. Since breakdown effects still occurs for the same
V_{BC} (although not visible in I_C but only in V_{BE}) as for forced I_B measurements, no
increase of the suitable extraction range is obtained. One can generally include a
sophisticated model for the breakdown including the temperature effects, but this is
considered here as being too error prone to maintain accurate extraction results.

The bias conditions explained above are summarized in Fig. 3.5. I_C obtained
from forced I_E is quite constant in a large region leading to a linear increase of ΔT.
However, (3.1) is not capable of calculating the correct value in the complete region.
Hence, using accurately extracted values for all model parameters describing I_T (cf.
2.4) would strongly increase the region for linear extrapolation but also strongly
increase the required extraction effort.

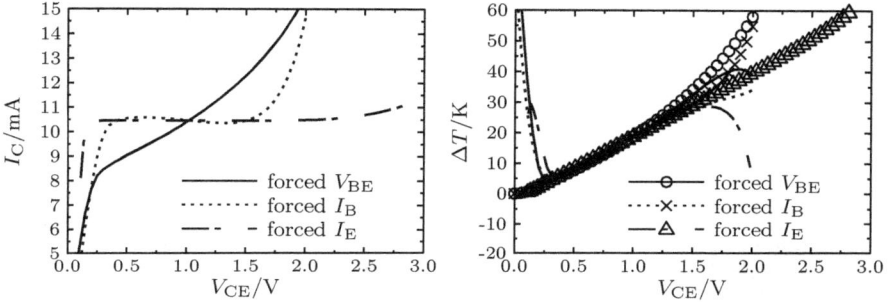

Figure 3.5: Comparison of I_C and ΔT obtained from measurements at different bias conditions. In the right picture the curves from calculating ΔT with (3.1) are added.

3.2.3 Extraction method

The extraction method can be written in mathematical representation as

$$R_{E,IB} - R_{E,VCE0} = 0, \tag{3.5}$$

where $R_{E,IB}$ is the value inserted in (3.2) to calculate ΔT with (3.1) and $R_{E,VCE0}$ the value calculated from (3.4). Thus, $R_{E,VCE0}$ can be considered as a non-linear function of $R_{E,IB}$, though a closed-form expression cannot be derived.

In this work, the above equation is solved by applying a numerical Newton method. Convergence is generally given by the fact that a lower value of $R_{E,IB}$ (with respect to the correct value) leads to an underestimation of ΔT and thus to larger values of $R_{E,VCE0}$ and vice-versa. Starting with an initial value ($R_E = 0\,\Omega$ can be used) numerical derivatives are calculated by a small change of the latter, e.g. $1\,\text{m}\Omega$. The iteration is terminated when either the change of $R_{E,VCE0}$ or the difference according to (3.5) is smaller than an user-defined value.

More details about the extraction procedure and sample values during the iteration are published in [PLS14]. There, different methods for solving (3.5) are also presented. It should finally be mentioned here that the extrapolation towards $\Delta T = 0\,\text{K}$ should always be performed using $\Delta T(V_{CE})/I_C(V_{CE})$ rather than only ΔT to reduce the impact of a possible non-linear $I_C(V_{CE})$.

3.2.4 Temperature dependence of the extracted values

Due to the extrapolation towards $\Delta T = 0\,\text{K}$ in this extraction method it is assumed that $R_{E,VCE0}$ can be considered for the actual ambient temperature. Since the value of $R_{E,IB}$ inserted in (3.2) during iteration depends on $R_{E,VCE0}$ and needs to be inserted for different T, an error is generated due to the neglected (since unknown) temperature dependence of R_E. Hence, results can be improved by extracting an initial temperature dependence of R_E and using this value during a repeated execution of the parameter extraction. However, the extraction should generally be performed at low self-heating in order to avoid the impact of changing values of R_E in (3.2).

3.2.5 Thermal resistance

After convergence of R_E, the value of the thermal resistance follows directly from (3.3). Different methods for extracting the value can be employed, including using the slope of ΔT vs. V_{CE} or bias (power) dependent values for each V_{CE}. The latter could be used as an utility for extracting a power dependent R_{th}. However, in this work no detailed investigation on this was performed.

Generally, the extracted value of R_{th} depends on the self-heating model applied, i.e. the calculation of the dissipated power. (3.3) assumed that only the power dissipated by the internal transistor directly affects the thermal conductivity in the transistor region, while the dissipated power due to voltage drops across series resistances is neglected. This corresponds to $flsh=1$ in the HICUM/L2 compact model. Obviously, since the extracted temperature increase is independent of the model for the power dissipation, the extracted value of R_{th} changes when including the voltage drops. Neglecting the base current, (3.3) and (3.4) change in this case to

$$\Delta T = R_{th} I_C V_{CE} \tag{3.6}$$

and

$$V_{CE,0} = 0, \tag{3.7}$$

respectively. The equation corresponds to $flsh=2$, except for the missing base current. Even in this case, the described extraction method for R_E can be applied, using (3.7) as function for optimization.

Using this equation, larger values for R_E than for $flsh=1$ are extracted, since larger values for ΔT are required for (3.7) in comparison to (3.4), for given I_B and V_{BE}, since this can only be achieved by larger R_E. It was found that the value of R_{th} was less affected by the self-heating model compared to R_E. Therefore, using one of the two optimization routines presented here results in different device temperatures when simulating the compact model. Effects on the base current are compensated by R_E but other parameters extracted at high currents are affected too. Thus, it is crucial to select one of the two models in advance of the parameter extraction process.

3.2.6 Collector resistance

So far, accurate values for R_{Cx} were assumed to be inserted in (3.4). Those can be obtained either from simple sheet resistance structures or from transistors connected as a tetrode as shown in [RKP+07]. Moreoever, extraction methods for single transistor characteristics can be applied. Following from (3.4), the principle of an extraction method for R_{Cx} base on electro-thermal effects can also be derived.

Provided measurements at two $I_{B,1/2}$ (for forced I_B measurements), two $V_{CE,1/2}$

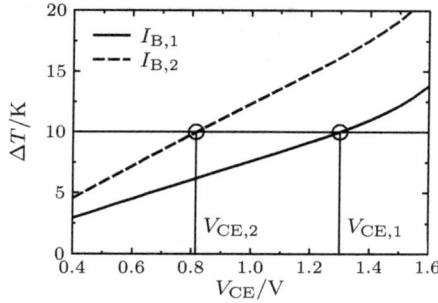

Figure 3.6: Visualization of $V_{CE,1}$ and $V_{CE,2}$ used for extraction of R_{Cx}. The sweeps were performed for $I_{B,1} = 10\,\mu A$ and $I_{B,2} = 20\,\mu A$.

with the corresponding $I_{C,1/2}$ and $I_{E,1/2}$ can be found, where the temperature increase is the same for both sweeps. The definition of both operating points is visualized in Fig. 3.6. Since the operating points are defined for the same temperature increase, all temperature dependent values are the same in both. Furthermore, the dissipated power is the same, allowing to write (using (3.4))

$$I_{C,1}\left(V_{CE,1} - I_{E,1}R_E - I_{C,1}R_{Cx}\right)R_{th} = I_{C,2}\left(V_{CE,2} - I_{E,2}R_E - I_{C,2}R_{Cx}\right)R_{th} \quad (3.8)$$

that is rewritten with respect to R_{Cx} into

$$R_{Cx} = \frac{I_{C,1}V_{CE,1} - I_{C,2}V_{CE,2} - R_E\left(I_{C,1}I_{E,1} - I_{C,2}I_{E,2}\right)}{I_{C,1}^2 - I_{C,2}^2}. \quad (3.9)$$

Note that R_{th} cancels out.

In this work, the above equation was applied by extracting R_E first with an initial value of R_{Cx}. The latter is then extracted by an outer iteration for R_{Cx}, thus repeating the R_E extraction with the updated value of R_{Cx}. Since even small errors in the calculation of ΔT will lead to completely wrong results for R_{Cx}, the temperature dependence of R_E is required to be included in the measurement. Also, the iteration for R_{Cx} should start with an overestimated value.

An example for the best accuracy possible with this method is given in Fig. 3.7. The results there are based on synthetic data and the R_{Cx} values inserted in the extraction were already the correct values. The results clearly show that the method is not applicable as it is. Further work is required but was omitted in this work.

A sensitivity analysis was performed for the extracted values of R_E and R_{th} depending on the value of R_{Cx} used during extraction and is presented in [PLS14]. The results presented there show only a weak influence of R_{Cx} on both R_E and R_{th} obtained by this method. For the latter, almost no influence exist, which is due to using the slope of $\Delta T = f(V_{CE})$. This slope is not affected by R_{Cx}.

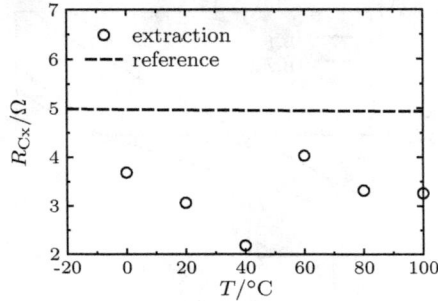

Figure 3.7: Extraction results for R_{Cx} for different temperatures. The extraction was performed on synthetic data. The reference model for R_{Cx} is included in the picture.

3.3 Extraction of the weight factors

Parameter extraction for the transfer current (2.125) is based on splitting the bias sweeps into regions dedicated to specific parts of the equation. A summary of the model parameters is given in Tab. 3.1. There, either the section describing its extraction or a reference for the extraction is given.

In the following, the application of the extraction methods is compared to both experimental results and a 1D simulation. The latter is used to compare the extracted value with the corresponding value from internal quantities to verify the method. The application of the extraction methods presented here is given in, e.g., [PSK+11] and for the given technology in this chapter and 4.4.

3.3.1 Low-bias region

The low-bias region is characterized by the classical Early effects. Here, the reverse Early effect is modeled by $h_{jEi}Q_{jEi}$, while the forward Early effect is taken into account by $h_{jCi}Q_{jCi}$ (cf. 2.4.3.1). Especially for parameter extraction it is assumed that effects in this region are dominated by both components. All effects caused by mobile charges are neglected.

In an initial extraction method published in [PSK09a], the extraction of the parameter h_{jEi} including its bias dependence was based on a numerical optimization on all parameters included in the model. Although quite accurate results were obtained, the method suffers from one fact: the parameters of $h_{jEi}(V_{B'E'})$ are strongly correlated when only using a small bias range for optimization. Thus, accurate results could be extracted but the physical nature of the actual values could be lost.

Therefore, the extraction was improved using the most promising method presented in [BCS+02] with results published in [PSK+11]. Using $V_{BC} = 0\,\text{V}$ and sufficiently low currents to avoid voltage drops across the series resistances, the transfer

Parameter	Description	Extraction method
c_{10}	GICCR constant	sec. 3.3.1
Q_{p0}	Zero-bias hole charge	from R_{sBi} tetrode-structures
h_{jEi0}	Weight factor for the BE depletion charge at $V_{B'E'} = 0\,V$, also refers to h_{jEi}	sec. 3.3.1
a_{hjEi}	Slope of the Ge-content in the base	sec. 3.3.1
r_{hjEi}	Smoothing parameter between low and medium injection	sec. 3.3.2
h_{jCi}	Weight factor for the BC depletion charge	cf. [BCS$^+$02]
h_{f0}	Weight factor for the low-bias minority charge	sec. 3.3.2
a_{hf0c}	Bias dependence of h_{f0}, cf. 2.136	sec. 3.3.2
h_{fE}	Weight factor for the high current emitter charge	sec. 3.3.3
h_{fC}	Weight factor for the high current collector charge	sec. 3.3.3

Table 3.1: Parameters describing the transfer current in HICUM/L2.

current reads[1]

$$I_{T,low} = \frac{c_{10}}{Q_{p0} + h_{jEi}Q_{jEi}} \exp\left(\frac{V_{B'E'}}{V_T}\right) \tag{3.10}$$

and can be rewritten into

$$\frac{\exp\left(\frac{V_{B'E'}}{V_T}\right)}{I_{T,low}} = \frac{Q_{p0}}{c_{10}} + \frac{h_{jEi}}{c_{10}}Q_{jEi}. \tag{3.11}$$

Here and in all following equations, the transfer current is replaced by the terminal collector current, which is justified in forward operating range, as long as no breakdown or saturation effects occur. The method in [BCS$^+$02] relates above equation to a linear equation of the form $y = mQ_{jEi} + n$, where the intercept point and slope are used to extract the parameters c_{10} and h_{jEi}, respectively. However, in case of a bias dependence of h_{jEi} (cf. section 2.4.2.4), no line is obtained when plotting above equation. This is shown by the curvature of the data with $h = 1$ in Fig. 3.8 for both 1D simulation and experimental data. Here, h is the normalized weight factor, see below. Although values for both can still be extracted from a linear fit, often a negative intercept point and, thus, a negative c_{10} is extracted. This effect is caused directly by the non-constant h_{jEi}.

For the extraction of the latter, the goal is to separate the extraction step of h_{jEi0} from (2.103) from that of a_{hjEi} and thus to avoid the strong coupling between

---

[1] The additional index "low" refers to the limitation of low injection for this formulation.

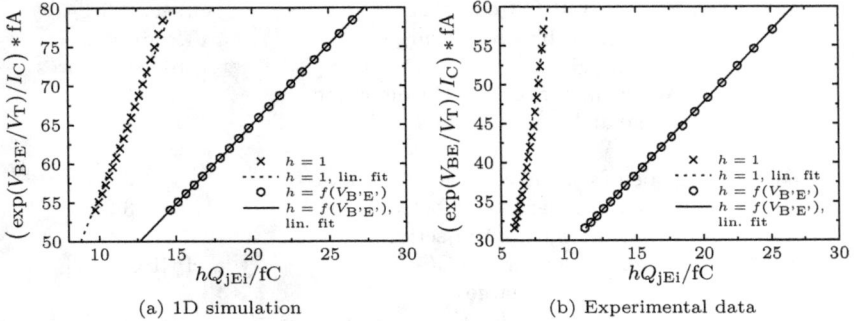

(a) 1D simulation   (b) Experimental data

Figure 3.8: Extraction of c_{10} and h_{jEi0} using eq. (3.11) and (3.16) for 1D simulation and experimental data at 27 °C. In this figure, symbols show extracted data, utilizing either a constant $h = 1$ or a bias dependent h with an previously extracted a_{hjEi}. The lines correspond to the application of (3.11).

both. The weight factor h_{jEi} is split into the zero-bias value and the bias dependent portion

$$h_{jEi} = h_{jEi0}h(V_{B'E'}). \tag{3.12}$$

The latter is only dependent on $V_{B'E'}$ and the parameter a_{hjEi} and is extracted by using a transformation of (3.10) into the normalized form

$$i_{T,low} = \frac{I_s}{1 + \frac{hv_{jEi}}{V_{Er}}} \exp\left(\frac{V_{B'E'}}{V_T}\right) \tag{3.13}$$

with

$$I_s = \frac{c_{10}}{Q_{p0}} \; , \; v_{jEi} = \frac{Q_{jEi}}{C_{jEi0}} \text{ and } V_{Er} = \frac{Q_{p0}}{h_{jEi0}C_{jEi0}}. \tag{3.14}$$

Using four operating points ($V_{BE,1..4}$; $I_{C,1..4}$) and the corresponding $h_{1..4}$ and $v_{jEi,1..4}$ allows writing

$$\frac{h_1 v_{jEi,1} - h_2 v_{jEi,2}}{h_3 v_{jEi,3} - h_4 v_{jEi,4}} = \frac{\dfrac{\exp(V_{BE,1}/V_T)}{I_{C,1}} - \dfrac{\exp(V_{BE,2}/V_T)}{I_{C,2}}}{\dfrac{\exp(V_{BE,3}/V_T)}{I_{C,3}} - \dfrac{\exp(V_{BE,4}/V_T)}{I_{C,4}}}. \tag{3.15}$$

By using this transformation, the unknown variables I_s and V_{Er} from (3.13) are canceled out, leaving a_{hjEi} from h as only variable. This nonlinear equation is solved iteratively for different sets of operating points, giving a set of values for a_{hjEi}, where the mean or median[1] value can be used. This allows to cancel out noise from limited

---

[1]Using the median is useful for very noisy results.

resolution of the measurement equipment.

A sample of extraction results is given in Fig. 3.9, showing extracted values for different voltage combinations and a large range of ambient temperatures. The strongly rounded shape for the experimental results is caused by the smoothing of $I_C(V_{BE})$, since even the smallest measurement noise can result in completely wrong results.

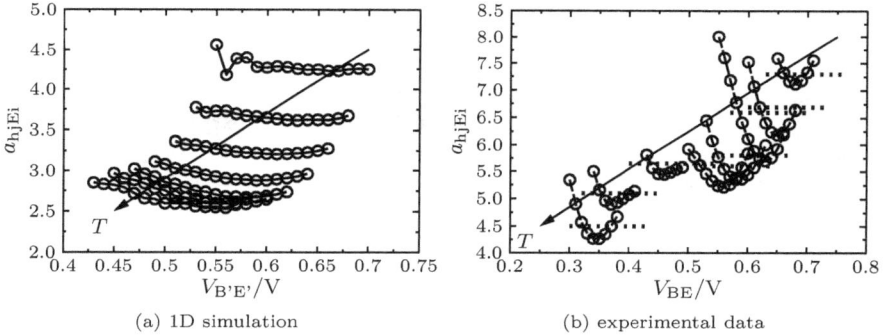

(a) 1D simulation   (b) experimental data

Figure 3.9: Extracted values of a_{hjEi} for different combinations of $V_{BE,1..4}$. Here, adjacent V_{BE} values were used, the circles always represent the smallest value. The arrows show the increase of the ambient temperatures. The temperature ranges in both cases are $T_{amb} = [-40\ldots125]\,°\mathrm{C}$. In (b) the thick dashed lines show the actual values of a_{hjEi} used for extracting the temperature parameters. Due to the weak curvature in (a), the smallest value of each curve was used.

A different method for extracting a_{hjEi} separately from h_{jEi0} was published in [SHD⁺12], rewriting the term $(\exp(u) - 1)/u$ in a form suitable to extract the parameter by use of the Lambert-W function. However, this method requires a known value of h_{jEi0} and therefore also only ends in numerical optimization of all parameters in a single loop.

Incorporating the extracted a_{hjEi}, eq. (3.11) is altered to

$$\frac{\exp\left(\frac{V_{BEi}}{V_T}\right)}{I_{T,low}} = \frac{Q_{p0}}{c_{10}} + \frac{h_{jEi0}}{c_{10}}hQ_{jEi}, \tag{3.16}$$

changing the abscissa to hQ_{jEi}. This finally leads to the desired linear form, allowing to extract both remaining parameters c_{10} and h_{jEi0} as shown in Fig. 3.8.

This extraction is performed in the operating region with low injection. Since this is accompanied with very low self-heating, the extraction of these three parameters is performed separately for a set of different ambient temperatures. Examples for $h_{jEi0}(T)$ are given in [PSK⁺11] and 4.3.4.2.

3.3.2 Medium bias region

This operating region is characterized by the beginning influence of the mobile charge. However, high current effects and, thus, the strong increase of the charge have not yet begun. The weighted charge in this region is dominated by $h_{f0}\tau_{f0}$. Using

$$Q_{pT,med} = \frac{c_{10}}{I_T} \exp\left(\frac{V_{B'E'}}{V_T}\right) - \left(Q_{p0} + h_{jEi}Q_{jEi} + h_{jCi}Q_{jCi}\right) \qquad (3.17)$$

allows to calculate h_{f0} as

$$h_{f0} = \frac{Q_{pT,med}}{I_{Tf}\tau_{f0}}. \qquad (3.18)$$

Note that depending on the current density in the extraction region, a calculation of $V_{B'E'}$ from V_{BE} might already be required using previously extracted values for the series resistances. Also, the actual device temperature will probably be increased with respect to the ambient temperature. Thus, calculating ΔT is a mandatory step before extracting h_{f0}.

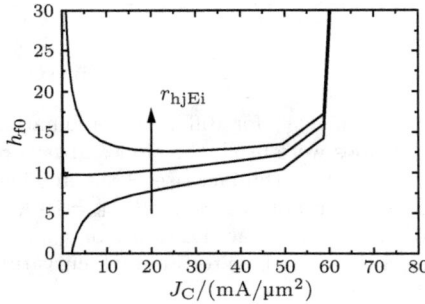

Figure 3.10: Value for h_{f0} according to (3.18) for 1D simulation data and $V_{B'C'} = 0\,\text{V}$ and $r_{hjEi} = [1, 3.2, 5]$.

Eq. (3.18) returns bias dependent value, with possibly very large positive or negative values for low currents. By fine-tuning r_{hjEi}, flat curves are obtained as demonstrated in Fig. 3.10.

In contrast to h_{jEi}, the extraction for h_{f0} is performed for several values of $V_{B'C'}$ to enable the extraction of the model parameter describing the $V_{B'C'}$ dependence in (2.136). Results for 1D data are shown in Fig. 2.47. For measurements, the values for different $V_{B'C'}$ result in different device temperatures due to self-heating. This is displayed in Fig. 3.11(a) by the decreasing values for high V_{BC} at high currents and in Fig. 3.11(b) by the step like curve of the extracted values. Using only the values for $V_{B'C'} = 0\,\text{V}$, ΔV_{gBE} is extracted and a correction of the extracted values is performed with

$$h_{f0,iso} = h_{f0} \exp\left(-\frac{\Delta V_{gBE}}{V_T}\left(\frac{T_{dev}}{T_{amb}} - 1\right)\right). \qquad (3.19)$$

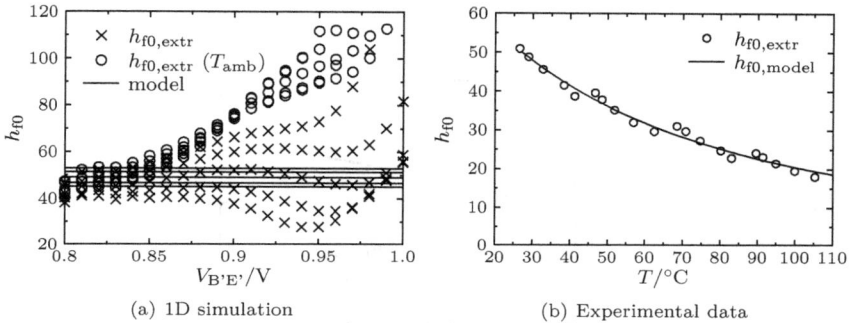

(a) 1D simulation

(b) Experimental data

Figure 3.11: Extraction of h_{f0} according to (3.18) for experimental data, $T = 27\,°C$ and $V_{BC} = [-0.5\ldots0.5]\,V$. Here, $h_{f0,\text{extr}}\ (T_{amb})$ represents the extracted values after performing a correction for the thermal effects.

Results for the model with respect to V_{BE} and V_{BC} are given in Fig. 3.11(a) and 3.12 showing the usability of the extraction method and verifying the physics-based models.

3.3.3 High current region

The extraction of h_{fE} and h_{fC} is performed similarly to that of h_{f0}, though with the additional subtraction of $h_{f0}I_{Tf}\tau_{f0}$ for the former and additionally $h_{fE}\Delta Q_{Ef}$ for the latter. Accurate models for the respective charges are required for the extraction. Especially the extraction of h_{fC} is still very error prone, if not impossible for certain technologies. A correct value of h_{fC} requires operation far in the high current region, where strong thermal effects and voltage drops across the series resistances occur. Due to thermal and avalanche breakdown, the current can often not be increased to values significantly larger than I_{CK}.

Typical curves for h_{fE} and h_{fC} extracted from experimental data are show in Fig. 3.13. As for h_{f0}, thermal effects are dominating the region in which both weight factors have the most influence on the transfer current. Thus, results for different V_{BC} can be used to increase the number of $h(T)$ pairs. The actual value for a given ambient temperature is calculated for both weight factors similarly to (3.19).

The extracted values for different ambient temperatures and V_{BC} are summarized in Fig. 3.14. Although no curve as perfect as for h_{f0} is extracted for this technology, the physical trend of decreasing values with increasing temperatures is still clearly visible. From the slope of $\log_{10}(h_{fE}) = f(1/T)$, the bandgap differences $v_{gB} - v_{gE}$ and $v_{gB} - v_{gC}$ are extracted and used for temperature scaling. The results of the corrected values are also included in Fig. 3.13. By using the temperature correction, the dependence on V_{BC} disappears for both weight factors as assumed from theory. For h_{fE}, almost constant values in the region where Q_{fE} dominates ΔQ_{fh} are extracted.

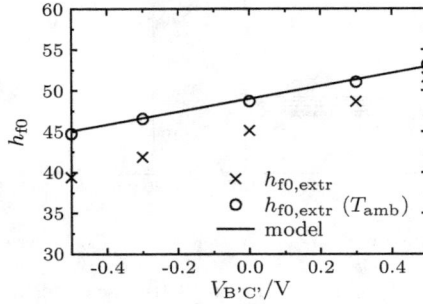

Figure 3.12: Model results for $h_{f0}(V_{B'C'})$ after performing the correction for the thermal effects.

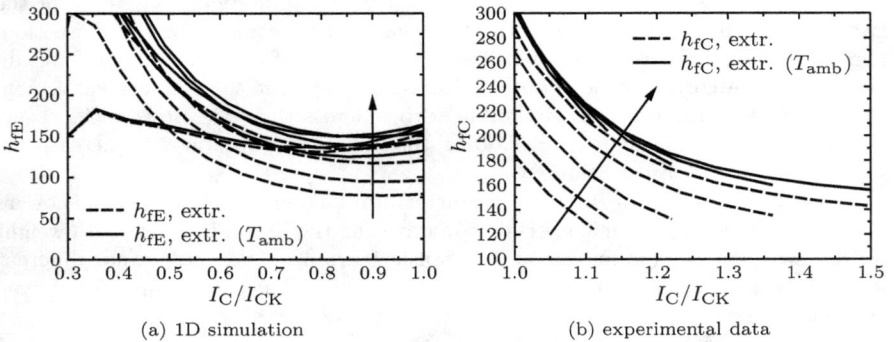

(a) 1D simulation

(b) experimental data

Figure 3.13: Extraction of h_{fE} and h_{fC} as a function of bias for experimental results and $T_{amb} = 27\,°C$. Here, $V_{BC} = [-0.5 \ldots 0.5]\,V$ with the arrow indicating increasing V_{BC} for the non-temperature scaled values.

Increasing values for low bias are caused by the same effect as explained for h_{f0}. Small deviations in the extracted weighted charge are amplified by the division by the charge. The same holds for h_{fC}, but the shape of this curve can also be explained by inaccuracies of the extraction of the charge or the charge model itself.

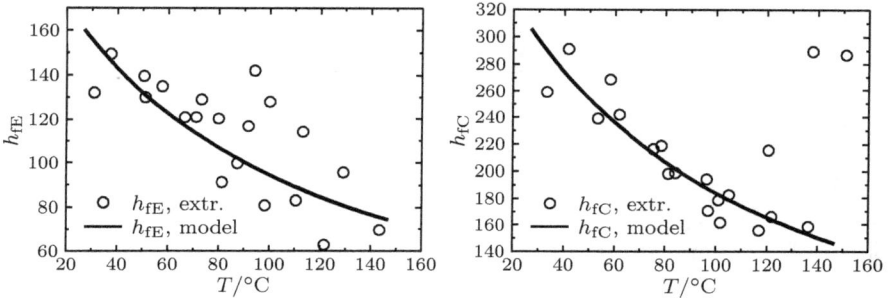

Figure 3.14: Extraction results for h_{fE} and h_{fC} for different ambient temperatures and V_{BC}. Model equations (2.142) and (2.143) are included.

Results given above are extracted from the selected technology in chapter 4. Sample results for a different technology were presented in [PSK$^+$11].

3.4 Geometry scaling

Geometry scaling in the context of this work refers to the calculation of model parameters that are dependent on the device geometry by applying dedicated scaling rules. Initial publications on scaling rules took the current gain ([GG76]) and the collector current ([Rei84]) into account. However, since charges correspond to dynamic currents, the same scaling rules can also be applied to those components. The often used principle for scaling currents of a transistor is to separate them into an area and a perimeter related component. In the following, the currents are generalized as an arbitrary device quantity Q. Thus, the scaling equation often applied reads

$$Q = \overline{Q_A} A_{E0} + Q' P_{E0}. \tag{3.20}$$

$A_{E0} = b_{E0} * l_{E0}$ is the actual emitter window area and $P_{E0} = 2(b_{E0} + l_{E0})$ the perimeter of the emitter. The index E0 refers to the actual dimensions, which are often different from the drawn ones.

Later publications ([SW96,SRR$^+$99]) not only improved the method by specifically adding corner rounding effects but also introduced scaling equations for minority charges and the onset of high current effects, including current spreading.

The extraction of the so-called process parameters $\overline{Q_A}$ and Q', i.e. the area and perimeter related component, is performed by a linear regression of Q/A_{E0} versus P_{E0}/A_{E0}. The method can be applied as long as a straight line is obtained for devices with different emitter lengths and widths.

In the course of this work, two specific geometry scaling issues are discussed. The first one discusses cases where (3.20) cannot be applied to experimental data anymore. The corresponding modeling is called here *General device scaling* and is discussed next. The other issue evaluated in this section covers the scaling of the transfer current in combination with the reverse Early effect.

3.4.1 General device scaling

3.4.1.1 Theory

In this work, a more general scaling approach than (3.20) is applied. The approach was introduced and discussed with respect to process specific issues in [Sch13]. Instead of separating a device quantity Q in an area and perimeter component, the following equation is used

$$Q = \overline{Q_A} A + Q'_b b + Q'_l l + Q_c, \tag{3.21}$$

where $A = b * l$. Here, b and l are the width and length used for scaling and are not limited to the emitter dimensions, although in most cases the latter are used. In contrast to the standard scaling, the perimeter component is separated into a width and length specific value, which are not necessarily equal. However, for the special case of $Q'_b = Q'_l$ and $Q_c = 0$, (3.21) simplifies to (3.20) with $Q' = Q'_b/2$. A graphical representation of (3.21) is given in Fig. 3.15.

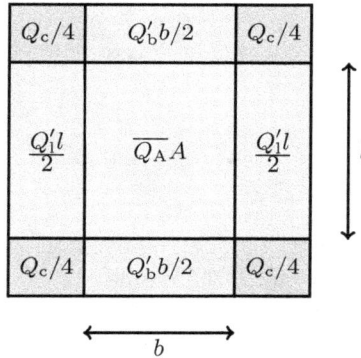

Figure 3.15: Sketch Q composed from the scaling parameters from (3.21).

The extraction of the components is based on a separate scaling with respect to b and l. In order to apply (3.21), a linear dependence of Q on b and l is required. In this case, Q is described by

$$Q = Q'_{b,l0}b + \Delta Q_{b,l0} \tag{3.22}$$

and

$$Q = Q'_{l,b0}l + \Delta Q_{l,b0}. \tag{3.23}$$

The indices l, b_0 and b, l_0 indicate that the scaling with respect to l is performed at a specific b_0 and vice versa. Since the scaling with respect to b is performed at a specific l_0, $Q'_{b,l0}$ is different from Q'_b and a function of l_0. Similarly, $\Delta Q_{b,l0}$ is different from Q_c and depends on l_0. The same holds for $Q'_{l,b0}$ and $\Delta Q_{l,b0}$ with respect to b_0. The scaling parameters from (3.21) are related to the extracted parameters of (3.22) and (3.23) by

$$\begin{bmatrix} l_0 & 1 & 0 & 0 \\ b_0 & 0 & 1 & 0 \\ 0 & 0 & l_0 & 1 \\ 0 & b_0 & 0 & 1 \end{bmatrix} \begin{bmatrix} \overline{Q_A} \\ Q'_b \\ Q'_l \\ Q_c \end{bmatrix} = \begin{bmatrix} Q'_{b,l0} \\ Q'_{l,b0} \\ \Delta Q_{b,l0} \\ \Delta Q_{l,b0} \end{bmatrix}. \tag{3.24}$$

These equations can be derived by formulas (3.22) and (3.23) with (3.21), respectively. However, since the matrix on the LHS of the above system of equations is singular, no unique solution for the scaling parameters can be obtained using above equations. A unique solution can only be obtained from a b-scaling at different l_0 and vice versa and, thus, requires the presences of matrix like geometry variations. That means that at least two devices with different b for at least two different l each need to be available, though a larger number is recommended. $\overline{Q_A}$ is then extracted from the slope of a linear regression of $Q'_{b,l0}$ versus different l_0 or from the slope of $Q'_{l,b0}$ versus different b_0, respectively. Using the same devices for both regressions, the same value is obtained from both. Q'_l and Q'_b are extracted from the intercept for both linear curves. Finally, Q_c is extracted from the intercept with the y-axis of the linear regression of $\Delta Q_{b,l0}$ versus l_0 or $\Delta Q_{l,b0}$ versus b_0.

3.4.1.2 Interpretation

The extracted values Q_b' and Q_l' can generally be caused by two phenomena. First, the perimeter component $Q' > 0$ of (3.20) results in $Q_b' > 0$ and $Q_l' > 0$ when using the correct dimensions. Second, wrongly inserted dimensions (e.g. due to effects during device processing) also yield perimeter components not equal to zero. In this case, these components are not actual physical quantities anymore. In [SMNJ04] it was discussed that non-standard scaling can occur simply due to (possibly different) offsets between the assumed dimensions and the actual ones. After obtaining and inserting the correct values of these offsets, ideal scaling (in terms of perimeter and area scaling) is obtained.

Assuming that the used values b and l for extraction differ from the actual values b_a and l_a by the offsets Δl and Δb such that

$$b = b_a + \Delta b \quad \text{and} \quad l = l_a + \Delta l \tag{3.25}$$

and that the actual value $Q_{c,a} = 0$, it follows for the extracted data

$$
\begin{aligned}
Q_b' &= Q_{b,a}' - \overline{Q_A}\Delta l \\
Q_l' &= Q_{l,a}' - \overline{Q_A}\Delta b \\
Q_c &= \overline{Q_A}\Delta b\Delta l - \left(Q_{b,a}'\Delta b + Q_{l,a}'\Delta l\right)
\end{aligned}
\tag{3.26}
$$

where $Q_{b,a}'$ and $Q_{l,a}'$ are the actually correct values. Thus, positive offsets lead to a reduction of Q_b' or Q_l' and even to negative values. The value and especially the sign of Q_c depends on the ratio of $\overline{Q_A}$ to both perimeter components and the sign of the offsets. If both offsets are negative, Q_c will always yield a positive value, compensating the missing area. For offsets with different signs, Q_c will generally become negative. If both offsets are large and positive, Q_c can again become positive, compensating the negative Q_b' and Q_l'. However, in case of constant offsets, i.e. both offsets are not dependent on b and l, the area related component is *not* affected.

Additional quantities providing information on the actual values are the intersects with the x-axis for $Q = 0$ from extraction versus b and l,

$$\Delta b_{l0} = -\Delta Q_{b,l0}/Q_{b,l0}' \quad \text{and} \quad \Delta l_{b0} = -\Delta Q_{l,b0}/Q_{l,b0}'. \tag{3.27}$$

Inserting (3.24) into (3.27) yields

$$\Delta b_{l0} = -\frac{Q_l'l_0 + Q_c}{\overline{Q_A}l_0 + Q_b'} \quad \text{and} \quad \Delta l_{b0} = -\frac{Q_b'b_0 + Q_c}{\overline{Q_A}b_0 + Q_l'}. \tag{3.28}$$

In case of existing offsets, they depend linearly on the latter parameter by

$$\Delta b_{l0} = \Delta b_{l0,a} + \Delta b$$
$$\Delta l_{b0} = \Delta l_{b0,a} + \Delta l. \tag{3.29}$$

Offsets from using the wrong dimensions during extraction directly affect these values. However, the extracted values clearly depend on l_0 and b_0 (in the common case Q'_b and $Q'_l \neq 0$) and can therefore not be directly used to calculate Δb and Δl. However, due to the direct influence, estimations are possible.

A discussion on corner rounding is only useful when assuming $Q'_b = Q'_l = 2Q'$, i.e. a constant perimeter component independent of the location. Other assumptions would lead to arbitrary results. Using this assumption leads to

$$Q_c = -(4 - \pi)r(\overline{Q_A}r + 2Q'), \tag{3.30}$$

where r is the radius of the corner rounding. This is the same result as given in [SRR$^+$99], though in a different context. Similar to the equations there, here also $r \leq b_{min}$ and $r \leq l_{min}$ is required. In case of corner rounding, a negative Q_c follows, since the overestimation by the assumed rectangular shape is compensated. The effects of (3.26) and (3.30) superimpose, which has implications when calculating the offsets as shown in 3.4.1.4. However, corner rounding has no influence on the extracted values of $\overline{Q_A}$, Q'_b and Q'_l, i.e. Q'.

3.4.1.3 Effective emitter area

For the application of scaling equations to the transfer current, a one-transistor model is preferred, since in compact models the calculation of the transfer current and related charges is one of the components with the largest impact on the runtime. A concept for enabling one-transistor models is defining an effective emitter area A_E so that (e.g. [Rei84])

$$\overline{I_A}A_E = \overline{I_A}A_{E0} + I'P_{E0}. \tag{3.31}$$

Here, A_{E0} and P_{E0} are calculated from b_{E0} and l_{E0}, the classical definition of which are the actual emitter dimensions. However, the application is not restricted to this definition, as long as the perimeter versus area scaling can be applied.

In the same way, an effective area can also be defined from (3.21). The value is then calculated by

$$A_E = A_{E0} + \frac{I'_b}{\overline{I_A}}b_{E0} + \frac{I'_l}{\overline{I_A}}l_{E0} + \frac{I_c}{\overline{I_A}}. \tag{3.32}$$

Above equation can be rewritten into

$$A_E = \left(b_{E0} + \frac{I'_l}{\overline{I_A}}\right)\left(l_{E0} + \frac{I'_b}{\overline{I_A}}\right) - \frac{I'_b I'_l}{\overline{I_A}^2} + \frac{I_c}{\overline{I_A}} = b_E l_E - \frac{I'_b I'_l}{\overline{I_A}^2} + \frac{I_c}{\overline{I_A}} \tag{3.33}$$

and thus results in a form similar to the classical scaling approach using γ_c as shown in, e.g., [SRR⁺99]. The effective emitter dimensions are hence calculated by

$$b_E = b_{E0} + \frac{I'_1}{\overline{I_A}} \quad \text{and} \quad l_E = l_{E0} + \frac{I'_b}{\overline{I_A}}. \tag{3.34}$$

The term $I'_b I'_1 / \overline{I_A}^2$ in (3.33) is the correction required when using (3.34) and $I_c / \overline{I_A}$ comprises all effects related to corner rounding.

3.4.1.4 Method demonstration

In the following, the extraction method described before is applied to synthetic data, i.e. model data with known results. Transistor geometries are defined for three width $b = [0.15, 0.30, 0.80]\,\mu m$ at three lengths $l = [2.5, 5.0, 10.0]\,\mu m$ each. The actual value Q is calculated by (3.20) with $\overline{Q_A} = 5/\mu m^2$ and $Q' = 0.2/\mu m$. A corner rounding with $r = 50\,nm$ is applied. During extraction, $\Delta b = 50\,nm$ and $\Delta l = 0.5\,\mu m$ is assumed according to (3.25).

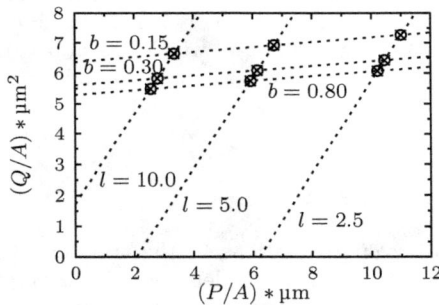

Figure 3.16: Results from the P/A scaling of the values obtained by (3.20) (circles) and results after extraction and application of (3.21) (crosses). Dashed lines represent linear curves obtained at fixed b_0 and l_0, respectively. Note, the slopes of the curves for different l_0 and b_0 are different.

The results for the P/A separation are shown in Fig. 3.16. Using the standard approach for the scaling leads to non-unique results, which are non-physical due to the negative area (and large perimeter) value obtained for fixed l_0. The parameter values after extraction are summarized in Tab. 3.2. While for $\Delta b = \Delta l = 0$, the correct results are obtained[1], completely different values are obtained when assuming incorrect dimensions during extraction. However, using the general scaling approach can accurately capture the effects of wrong dimensions during extraction and offsets due to, e.g., corner rounding. However, the parameters may lose their physical correlation.

---

[1]The factor 2 is caused by the definition of Q' in (3.20) and Q'_b and Q'_1 in (3.21).

Parameter	no offsets	offsets
$Q_A * \mu m^2$	5	5
$Q'_b * \mu m$	0.4	2.9
$Q'_l * \mu m$	0.4	0.15
Q_c	-0.0279	0.0271

Table 3.2: "No offsets" represents the extraction with $\Delta b = \Delta l = 0$, while the column "offsets" shows the results with the assumed offsets.

Using the assumption[1] $Q_{c,a} = 0$ and $Q'_{b,a} = Q'_{l,a} = 2Q'$, one can solve (3.26) with respect to Q' (replacing Q'_b and Q'_l), Δb and Δl. For the given data, the solution is visualized in Fig. 3.17. As shown in the top view, more than one solution is possible, though only one solution provides $Q' < 0$. However, when assuming the actual $Q_{c,a} = 0$, i.e. no corner rounding is present, the obtained solution is not correct, though with only small deviations.

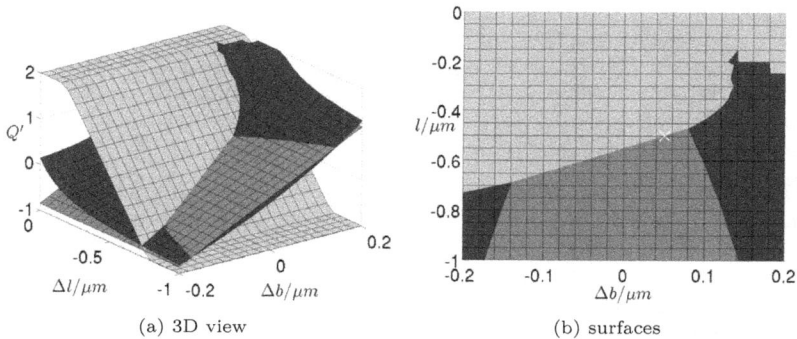

(a) 3D view

(b) surfaces

Figure 3.17: Graphical representation of the solution of (3.26) with respect to Q', Δb and Δl for synthetic data. $Q_c = 0$ is assumed. The correct solution is marked with a white cross. The light gray surface represents the first, the medium gray the second and the dark gray curve the third equation in (3.26).

Assuming the correct corner rounding radius r which, however, can only be obtained from pictures of the final emitter window, the correct solutions are obtained as shown in Fig. 3.18. Using this method, one can obtain values for Δb and Δl, provided the assumed physics-based assumptions are valid for the given technology and quantity.

---

[1]This is the general assumption for the transfer current, although corner rounding needs to be taken into account according to (3.30).

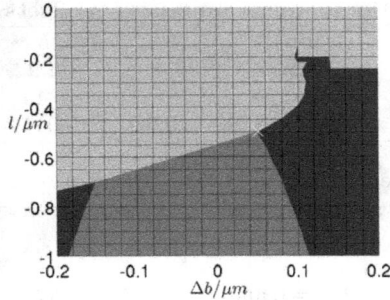

Figure 3.18: Solution of (3.26) when taking the correct r for corner rounding into account. The correct solution is marked with a white cross.

3.4.2 Bias dependent low injection scaling

Note that in this chapter, the indices "a" and "p" (e.g. V_{Era} and V_{Erp}) are used for the area and perimeter related component, respectively. The index "i" (e.g. V_{Eri}) represents the actual value of a parameter for a specific geometry. A parameter without any of the former post-subscript (e.g. V_{Er}) is used for equations that can be applied for all variations.

The concept of γ_c for calculating the effective area has implications for possibly different parameters describing the reverse Early effect of the internal and peripheral transistor. This is the general case that is often caused by different doping profiles in the pn-junction below the emitter window and the spacer. In early publications, γ_c was already modeled as a function of $V_{B'E'}$ by using the non-ideality factors of both current components ([Rei84]). For more recent technologies, this effect was also reported ([Cel09]). In this presentation, a more detailed model of γ_c was derived by using the different reverse Early voltages. For the final transfer current, the bias dependence of the effective emitter area was included into the model equation. The drawback of this approach is the requirement of adding geometry related parameters into the model.

In both works, a zeros bias value γ_{c0} is (or can easily be) defined so that

$$\gamma_c = \gamma_{c0} f_c (V_{B'E'}) \tag{3.35}$$

with $f_c (V_{B'E'})$ describing the bias dependence. Since $f_c \neq 1$ is caused by different parameters of the reverse Early effect for the internal and peripheral transistor as well as (for large bias) differently scaling series resistances, here the effective area is calculated only using γ_{c0}. The effect of f_c is incorporated into the final model parameters. The transfer current equation of HICUM/L2 is rewritten into the normalized form (3.13) that is valid for low injection and $V_{B'C'} = 0\,\text{V}$ and used for the area and perimeter current component separately, yielding

$$I_{\mathrm{Ta}} = \frac{\overline{J_{\mathrm{s,A}}} A_{\mathrm{E0}}}{1 + \frac{h_a v_{\mathrm{jEa}}}{V_{\mathrm{Era}}}} \exp\left(\frac{V_{\mathrm{B'E'}}}{V_{\mathrm{T}}}\right) \quad \text{and} \quad I_{\mathrm{Tp}} = \frac{I_s' P_{\mathrm{E0}}}{1 + \frac{h_p v_{\mathrm{jEp}}}{V_{\mathrm{Erp}}}} \exp\left(\frac{V_{\mathrm{B'E'}}}{V_{\mathrm{T}}}\right) \tag{3.36}$$

with the Early voltages and normalized charges for the internal and peripheral current $V_{\mathrm{Er(a,p)}}$ and $v_{\mathrm{jE(a,p)}}$, respectively. h_a and h_p are values of the weight factors $h_{\mathrm{jE(a,p)}}$ for the internal and peripheral transistors the normalized to the zero-volt value (cf. (3.12)). For deriving a scaling equation for V_{Er}, the limit of the normalized current for $V_{\mathrm{B'E'}} \to 0$ is calculated. For (3.13) it reads

$$\lim_{V_{\mathrm{B'E'}} \to 0} \left(\frac{I_s \exp\left(V_{\mathrm{B'E'}}/V_{\mathrm{T}}\right)}{I_{\mathrm{T}}\left(V_{\mathrm{B'E'}}\right)} - 1\right) \frac{1}{v_{\mathrm{jEi}}} = \lim_{V_{\mathrm{B'E'}} \to 0} \frac{h}{V_{\mathrm{Er}}} = \frac{1}{V_{\mathrm{Er}}}. \tag{3.37}$$

Although due to different doping profiles generally v_{jEa} is not equal to v_{jEp}, $v_{\mathrm{jEa}} = v_{\mathrm{jEp}} = v_{\mathrm{jEi}}$ is assumed in order to apply above equation. Using γ_{c0} for calculating A_{E}, the saturation current for the scaled model is $\overline{J_{\mathrm{s,A}}} A_{\mathrm{E}} = \overline{J_{\mathrm{s,A}}} A_{\mathrm{E0}} + I_s' P_{\mathrm{E0}}$. Also, the current measured at the terminals is the sum of both components. (3.37) is applied to (3.20) yielding

$$\lim_{V_{\mathrm{B'E'}} \to 0} \left(\frac{\overline{J_{\mathrm{s,A}}} A_{\mathrm{E}}}{\frac{\overline{J_{\mathrm{s,A}}} A_{\mathrm{E0}}}{1 + h_a v_{\mathrm{jEi}}/V_{\mathrm{Era}}} + \frac{I_s' P_{\mathrm{E0}}}{1 + h_p v_{\mathrm{jEi}}/V_{\mathrm{Erp}}}} - 1\right) \frac{1}{v_{\mathrm{jEi}}} =$$

$$\lim_{V_{\mathrm{B'E'}} \to 0} \left(\frac{\overline{J_{\mathrm{s,A}}} A_{\mathrm{E}} \left(1 + \frac{h_a v_{\mathrm{jEi}}}{V_{\mathrm{Era}}}\right)\left(1 + \frac{h_p v_{\mathrm{jEi}}}{V_{\mathrm{Erp}}}\right)}{\overline{J_{\mathrm{s,A}}} A_{\mathrm{E0}} \left(1 + \frac{h_p v_{\mathrm{jEi}}}{V_{\mathrm{Er}}}\right) + I_s' P_{\mathrm{E0}} \left(1 + \frac{h_a v_{\mathrm{jEi}}}{V_{\mathrm{Era}}}\right)} - 1\right) \frac{1}{v_{\mathrm{jEi}}}. \tag{3.38}$$

The usage of the limit rather than the value for a specific $V_{\mathrm{B'E'}}$ allows to simplify above equation. Since

$$\lim_{V_{\mathrm{B'E'}} \to 0} v_{\mathrm{jE}} = 0 \quad \text{and} \quad \lim_{V_{\mathrm{B'E'}} \to 0} \frac{\mathrm{d} v_{\mathrm{jE}}}{\mathrm{d} V_{\mathrm{B'E'}}} = 1, \tag{3.39}$$

the argument in the RHS of (3.38) is rewritten into

$$\lim_{V_{\mathrm{B'E'}} \to 0} \left(\frac{\overline{J_{\mathrm{s,A}}} A_{\mathrm{E}} \left(1 + \frac{h_a v_{\mathrm{jEi}}}{V_{\mathrm{Era}}}\right)\left(1 + \frac{h_p v_{\mathrm{jEi}}}{V_{\mathrm{Erp}}}\right)}{\overline{J_{\mathrm{s,A}}} A_{\mathrm{E0}} \left(1 + \frac{h_p v_{\mathrm{jEi}}}{V_{\mathrm{Er}}}\right) + I_s' P_{\mathrm{E0}} \left(1 + \frac{h_a v_{\mathrm{jEi}}}{V_{\mathrm{Era}}}\right)} - 1\right) \frac{1}{v_{\mathrm{jEi}}} =$$

$$\frac{\overline{J_{\mathrm{s,A}}} A_{\mathrm{E}} \left(\frac{1}{V_{\mathrm{Era}}} + \frac{1}{V_{\mathrm{Erp}}}\right)\left(\overline{J_{\mathrm{s,A}}} A_{\mathrm{E0}} + I_s' P_{\mathrm{E0}}\right) - \overline{J_{\mathrm{s,A}}} A_{\mathrm{E}} \left(\overline{J_{\mathrm{s,A}}} A_{\mathrm{E0}} \frac{1}{V_{\mathrm{Erp}}} + I_s' P_{\mathrm{E0}} \frac{1}{V_{\mathrm{Era}}}\right)}{\left(\overline{J_{\mathrm{s,A}}} A_{\mathrm{E0}} + I_s' P_{\mathrm{E0}}\right)^2}. \tag{3.40}$$

Note that the different values of the derivative $\mathrm{d}h/\mathrm{d}V_{\mathrm{B'E'}}$ at $V_{\mathrm{B'E'}} \to 0$ are neglected.

The final equation for V_{Eri} reads

$$V_{\text{Eri}} = \frac{\overline{J_{s,A}} A_E}{\overline{J_{s,A}} A_E \left(\frac{1}{V_{\text{Era}}} + \frac{1}{V_{\text{Erp}}} \right) - \left(\overline{J_{s,A}} A_{E0} \frac{1}{V_{\text{Erp}}} + I'_s P_{E0} \frac{1}{V_{\text{Era}}} \right)}. \tag{3.41}$$

From (3.14), the actual value of h_{jEi} is calculated.

Using a similar approach, the scaling of a_{hjEi} is derived. The limit of h reads

$$\lim_{V_{\text{B'E'}} \to 0} \frac{\mathrm{d}h}{\mathrm{d}V_{\text{B'E'}}} = \frac{a_{\text{hjE}}}{2} \frac{z_E}{V_{\text{dE}}} \tag{3.42}$$

Calculating this limit for the internal and peripheral transistor as well as for the scaled current allows to write

$$\lim_{V_{\text{B'E'}} \to 0} \frac{\mathrm{d}h_{\text{jEi}}}{\mathrm{d}V_{\text{B'E'}}} = \frac{1}{\overline{J_{s,A}} A_E} \left[\overline{J_{s,A}} A_E \left(a_{\text{hjEa}} \frac{z_{\text{Ea}}}{V_{\text{dEa}}} + a_{\text{hjEp}} \frac{z_{\text{Ep}}}{V_{\text{dEp}}} \right) \right.$$
$$\left. - \left(\overline{J_{s,A}} A_{E0} a_{\text{hjEp}} \frac{z_{\text{Ep}}}{V_{\text{dEp}}} + I'_s P_{E0} a_{\text{hjEa}} \frac{z_{\text{Ea}}}{V_{\text{dEa}}} \right) \right] = a_{\text{hjEi}} \frac{z_{\text{Ei}}}{V_{\text{dEi}}}, \tag{3.43}$$

which is rewritten with respect to a_{hjEi}. Note that no assumptions with respect to v_{jE}, i.e. z_E and V_{dE} are required here. The specific values for the internal and peripheral transistor are included. However, when assuming $v_{\text{jEa}} = v_{\text{jEp}} = v_{\text{jEi}}$, above equation simplifies to

$$a_{\text{hjEi}} = \frac{\overline{J_{s,A}} A_E (a_{\text{hjEa}} + a_{\text{hjEp}}) - \left(\overline{J_{s,A}} A_{E0} a_{\text{hjEp}} + I'_s P_{E0} a_{\text{hjEa}} \right)}{\overline{J_{s,A}} A_E}. \tag{3.44}$$

The extraction of the process specific values V_{Era} and V_{Erp} as well as a_{hjEa} and a_{hjEp} is either performed directly from $\overline{J_A}(V_{\text{B'E'}})$ and $I'(V_{\text{B'E'}})$, i.e. the result from the perimeter over area scaling at different (low bias) $V_{\text{B'E'}}$, or from an optimization of V_{Eri} and a_{hjEi} extracted from transistors with different geometries. The extraction of the latter is described in 3.3.1. The Early voltage is extracted similarly to h_{jEi} (also shown in 3.3.1) by a linear regression of

$$\frac{\exp\left(V_{\text{B'E'}}/V_T\right)}{I_C} = f(h * v_{\text{jE}}). \tag{3.45}$$

From the intercept with the y-axis, the saturation current is obtained, while the slope results in the Early voltage. An application of (3.41) and (3.43) is given in Fig. 3.19. In this example, $\overline{J_A}(V_{\text{B'E'}})$ and $I'(V_{\text{B'E'}})$ with given V_{Era} and V_{Erp} were used to calculate the transfer currents for different geometries. Based on those, an extraction of V_{Eri} and a_{hjEi} was performed. The extraction of V_{Eri} was performed in a very low bias region and a medium bias region. While perfect agreement was obtained for the former, small deviations occur for the much more practical latter case. In order to show the influence of $h(V_{\text{B'E'}})$ an application of (3.41) was performed on data with both a constant and a bias dependent value of h. In both cases, the scaling equation provides an accurate estimation. Applying (3.43) to the extraction results also provides very good agreement.

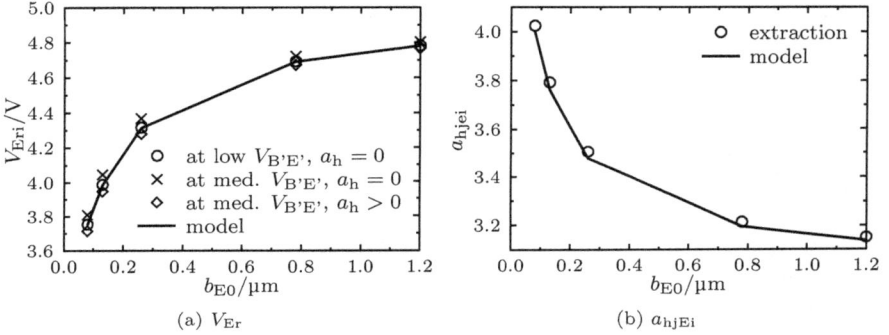

(a) V_{Er}

(b) a_{hjEi}

Figure 3.19: Application of (3.41) and (3.43) to synthetical data for different b_{E0}. In the left plot, the extraction was performed with $h_{a,p} = 1$ for $V_{B'E'} < 0.1\,V$ ("low $V_{B'E'}$") and for $0.4\,V < V_{B'E'} < 0.7\,V$ ("med. $V_{B'E'}$") and for the latter range also with $h_{a,p} = f(V_{B'E'})$. Here, $V_{Era} = 5\,V$ and $V_{Erp} = 3\,V$ as well as (for $a_h > 0$) $a_{hjEa} = 3$ and $a_{hjEp} = 5$ were used.

If the general scaling is applied rather than the classical scaling, (3.41) can easily be adopted to (3.21). Provided all components in (3.21) can be described by equations as in (3.36)[1], the same approach as in (3.38) is used, leading to

$$\frac{1}{V_{Eri}} = \sum_{a,b,l,c} \frac{1}{V_{Er}} - \frac{1}{\overline{J_{s,A}} A_E} \left(\overline{J_{s,A}} A_{E0} \sum_{b,l,c} \frac{1}{V_{Er}} + I'_{s,b} b_{E0} \sum_{a,l,c} \frac{1}{V_{Er}} \right.$$

$$\left. + I'_{s,l} l_{E0} \sum_{a,b,c} \frac{1}{V_{Er}} + I_{s,c} \sum_{a,b,l} \frac{1}{V_{Er}} \right). \tag{3.46}$$

with

$$\sum_{a,b,l,c} \frac{1}{V_{Er}} = \frac{1}{V_{Era}} + \frac{1}{V_{Erb}} + \frac{1}{V_{Erl}} + \frac{1}{V_{Erc}}. \tag{3.47}$$

Similar, the general scaling equation for a_{hjEi} can be derived based on (3.43).

---

[1] The parameters for the components are now defined by the indices "b" and "l" for the width and length dependent component and "c" for the constant component.

4 | Characterization of advanced SiGe HBTs

4.1 Description of the technology

The parameter extraction and model results presented in this chapter are from the SG13G2 technology from IHP. This SiGe:C high-speed technology offers an operating speed of $(f_T/f_{max}) = (300/500)$GHz and is discussed here very briefly. The HBT process was developed during Dotfive ([DVHH+09a,DVHH+09b,CMH+11,DOT11]) and specifically introduced in [HBB+10]. The final integration of the HBTs into a BiCMOS technology, leading to the actual SG13G2 technology, is described in [RHF12,RH12].

The key differences of this technology to conventional high-speed SiGe technologies are described in [HRB+02,HBK+06]. The whole transistor is fabricated in a single active area. This region is surrounded by a shallow trench isolation (STI). In contrast to conventional technologies, no trench isolation between the collector and emitter contacts exist. This strongly reduces the collector sheet resistance and avoids the usage of sinkers. A schematic cross-section of a transistor from this technology is shown in Fig. 4.1. While the highly doped collector is implanted at the surface of the wafer and surrounded by a STI, the base is epitaxially grown on the opened collector area in the SiO$_2$. This finally yields a mesa-type structure. Additionally, as shown in the figure, an elevated extrinsic base ([RHB+03]) is grown to reduce the base resistance.

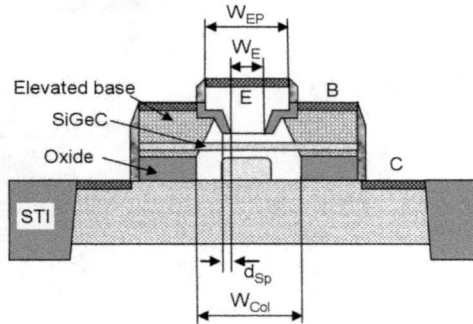

Figure 4.1: Schematic cross-section of a transistor from SG13G2, adopted from [RHF12].

4.2 Test chip description

For the characterization and parameter extraction of the selected technology a test chip was designed containing test structures as described below. A screenshot of the layout is presented in Fig. 4.2 representing a total chip area of $3.2 \text{x} 3 \text{ mm}^2$. Except for two small circuits, all structures implemented on the chip are dedicated to parameter extraction and model verification. Among those, transistors in different geometries and contact configurations as well as special structures for certain process parameters are available.

Figure 4.2: Screenshot of the layout from the complete test chip used for the device characterization in this work.

The structures designed for this test chip follow closely the content of [SKAA13]. In 4.2.1 and 4.2.2 a short summary of the implemented structures is given. A more comprehensive list is available in [Paw13]. A notable difference of this design compared to earlier designs is the implementation of devices in both 100 µm and 75 µm pads. However, only a few devices are available for the latter. This allows the characterization of devices in higher frequency bands, e.g. [220 ... 325] GHz, where probes generally have a maximum pitch of 75 µm. Furthermore, some designated reference devices are available more than once, allowing further characterization on the same die once some pads of the reference devices are worn out.

4.2.1 RF devices

4.2.1.1 Transistors

The majority of RF devices are transistors in different geometries and contact configurations. For the characterization of the given process two different kinds of devices are distinguished. On the one hand, devices are designed in a CEB configuration, where the base and collector contacts are at the fore-sides of the emitter window. These devices are labeled as *PDK configuration*, because this is the only contact configuration that is available in the official PDK. Furthermore, these devices are only fabricated with fixed emitter dimensions, thus, they are not suitable for a scaled extraction. In the PDK a scaling using a different number of parallel devices (N_x) is possible. Therefore, devices with different $N_x=[1, 2, 4, 8]$ were implemented. All parallel devices share a common substrate ring. In the following sections these devices are labeled as PDK_Nx_x where x is the number of parallel devices.

Configuration	b_E	l_E			
		1.10 µm	2.66 µm	5.16 µm	10.16 µm
CEB	0.13 µm	x		x	
CBE	0.13 µm	x		x	
CBEB	0.13 µm	x		x	
CBEBC	0.13 µm	x	x	x	x
	0.18 µm				x
	0.26 µm		x	x	x
	0.39 µm				x
	0.78 µm		x	x	x
(CBEBC)$_2$	0.13 µm				x
(CBEBC)$_4$	0.13 µm				x
CB(EB)$_2$C	0.13 µm			x	x

Table 4.1: Available geometries of the emitter window and corresponding contact configurations. Numbers in indices mark repetition of the preceding part in parentheses, e.g. CB(EB)$_2$C means CBEBEBC.

For a scaled extraction devices in a *standard* contact configuration were also designed, i.e. device with base and collector contacts at the sides of the emitter window. Although the geometry variations were performed using symmetrical CBEBC devices, different configurations were also designed. Tab. 4.1 provides a comprehensive list. In the following sections these devices will be referred to as M_CONFIG_bEx1E with the number of parallel devices M, the contact configuration CONFIG and the emitter dimensions bE and 1E, e.g. 1x_CBEBC_0p13x10p16. As with the PDK devices the original layouts were adopted from [Fox10] and adjusted to the required geometries and contact configurations. A comparison of the general layout differences between the PDK and standard configuration is given in Fig. 4.3. Here, only those layers are compared which are officially provided by IHP for the PDK devices. Although the CBEBC transistors are not available in the PDK and are, thus, not representative for the process[1], for these also only a few layers are displayed. The same holds for special test structures (cf. Fig. 4.9).

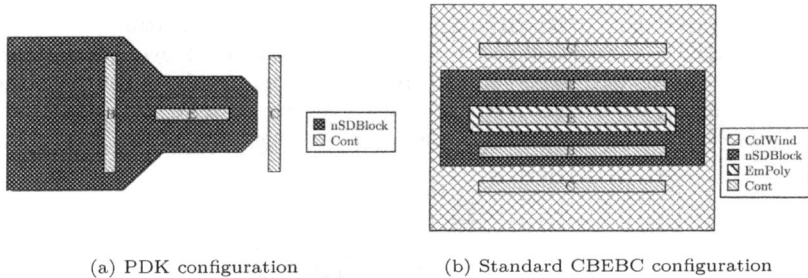

(a) PDK configuration   (b) Standard CBEBC configuration

Figure 4.3: Partial layouts for a device in PDK configuration and a device in standard CBEBC configuration. Only a few selected layers are shown. Dimensions in both pictures are not drawn to scale.

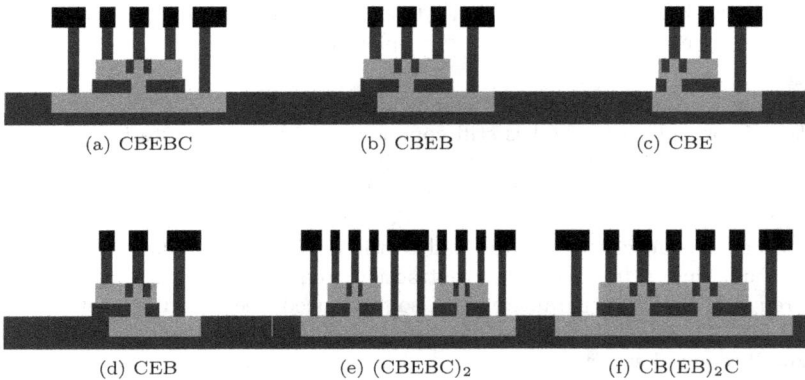

(a) CBEBC   (b) CBEB   (c) CBE

(d) CEB   (e) $(CBEBC)_2$ (f) $CB(EB)_2C$

Figure 4.4: Simplified cross-sections of transistors with different contact configurations.

In order to visualize the differences between the contact configurations cross-sections of devices according to Tab. 4.1 are given in Fig. 4.4. Note that for devices in $(CBEBC)_n$ configurations two parallel collector contacts are used between two

---

[1]Some of the critical dimensions are different between the PDK and the CBEBC devices.

devices in order to provide a better scaling of the external collector resistance and to reduce the effect of electromigration.

4.2.1.2 Deembedding structures

The interconnection of a transistor is shown in Fig. 4.5. The general design consists of a connection to the base and collector in the lowest metal layer (*Metal1*) which is connected via the next layer (*Metal2*) to all overlying metals outside of the active area. The emitter connection is realized by *Metal2* and connected to the device via Metal1 directly above the emitter contact.

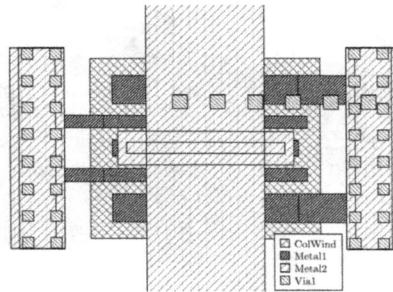

Figure 4.5: Interconnection of a transistor the metal lines. Shown is a transistor with $A_E = 0.13x2.66\,\mu m^2$. Only the lowest two metal layers and the corresponding via layer are displayed.

Standard two-step open-short deembedding (e.g. [WCW87,CB91]) is applied to the transistors. This method is still commonly used for characterizing high-speed bipolar devices (e.g. [HJJ$^+$07]). Although it was shown in [THJB05] that standard open-short deembedding fails for high-speed integrated inductors and frequencies higher than roughly 20 GHz, the method described in this publication is not applied here, because connection from the pad to the device is much simpler for this test chip. Furthermore, for parameter extraction frequencies larger than 20 GHz are only required for NQS parameters. Also, advanced deembedding methods as in, e.g., [LCN$^+$03] were not applied, because not enough chip space was available to design all required structures.

The deembedding structures of the device from Fig. 4.5 are given in Fig. 4.6. The open-structure is defined down to *Metal2*. Thus, no direct device interconnection is included. This structure allows to deembed the capacitance from the pads as well as the wide metal lines. The interconnects of the device are included in the short-structure where all contacts are connected onto the lowest metal layer by a low ohmic area.

Deembedding structures were only designed for CBEBC devices with different l_E. The same structures were used for different b_E due to the small differences in the interconnects for width scaled devices. Furthermore, for each PDK device with

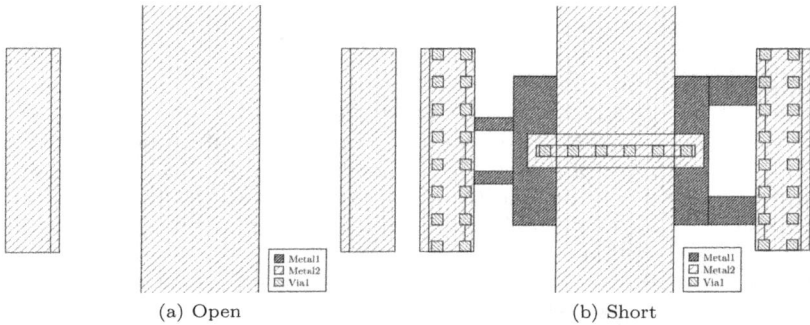

(a) Open   (b) Short

Figure 4.6: Interconnection of the deembedding structures as the same device as in Fig. 4.5.

a different N_{x} dedicated deembedding structures were designed as well as for devices with variations in the contact configuration. However, for the latter only a minimized set was designed (cf. [Paw13]). For future test chips, missing deembedding structures can be characterized either by EM simulations and/or scaling based on elaborate equivalent circuits (e.g. [LYS14]).

4.2.1.3 DC deembedding

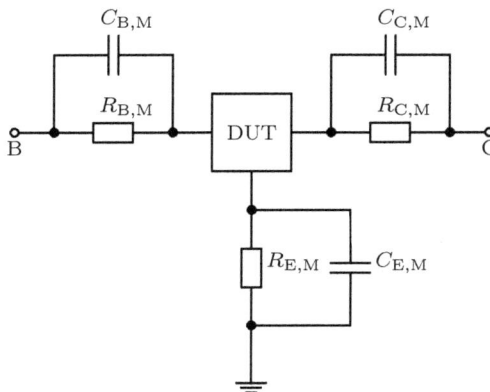

Figure 4.7: Circuit used for compact model simulations. Very large capacitance values in parallel to the resistances are used in order to remove any impact on small-signal results.

As discussed in previously in chapter 4.2.1.2, open-short deembedding is applied to the RF-measurements. However, in order to avoid inconsistent data between DC operating point and corresponding small-signal parameters a so-called *DC deembedding* is performed based on DC measurements of the deembedding short.

The electrical circuit used for compact model simulations is given in Fig. 4.7. Resistances caused by metal interconnects ($R_{B,M}$, $R_{C,M}$ and $R_{E,M}$) are connected in series to the DUT, since these resistances are not part of the device and hence the compact model. For RF data, however, the standard open-short deembedding removes the impact of these resistances. Therefore large capacitances are added in parallel during simulation.

In the DC case the deembedding short is characterized by the equivalent circuits given in Fig. 4.8. The physically correct representation is shown in (a), containing the same resistances as in Fig. 4.7. Based on standard data from standard measurement setups the resistances from (b) can be extracted more easily. Using the *wye-delta-transformation*, the resistances can be transformed from one circuit into another, allowing to easily characterize the resistances from metal interconnects.

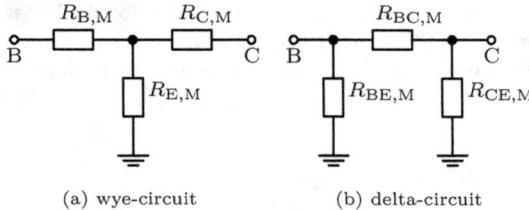

(a) wye-circuit   (b) delta-circuit

Figure 4.8: Circuit of the deembedding short in *wye* and *delta* layout.

A short overview over the extracted data for a few devices is given in Fig. 4.2. For the PDK devices, $R_{B,M}$ and $R_{C,M}$ roughly scale with the number of parallel devices, while R_E is very small. Since R_B and R_{Cx} from the devices are expected to scale with N_x as well, the impact would be the same for all devices. On the other hand, the resistances from the connection at base and collector side are nearly constant for the CBEBC devices, while $R_{E,M}$ scales almost perfectly. However, since R_B and R_{Cx} scale with $1/l_E$, not taking $R_{B,M}$ and $R_{C,M}$ into account would cause an increasing error for very large devices.

4.2.2 DC structures

DC structures on this test chip can be separated into three categories,

- transistors with four contacts, allowing a separation of emitter and substrate currents,

- structures for obtaining the sheet resistance of specific areas of the transistor and

Device	RB,M	RC,M	RE,M
PDK_Nx_1	4.3654	4.612	0.058443
PDK_Nx_2	2.7613	2.8923	0.053982
PDK_Nx_4	1.8558	1.74	0.085277
PDK_Nx_8	1.3711	1.1458	0.061009
1x_CBEBC_0p13x1p10	3.1024	3.6528	6.206
1x_CBEBC_0p13x2p66	3.3158	3.4575	2.1249
1x_CBEBC_0p13x5p16	3.3651	3.4434	1.1145
1x_CBEBC_0p13x10p16	3.4691	3.5185	0.64279

Table 4.2: Extracted parameters for the resistances of the metal interconnects for a number of selected devices.

- test structures and transistors for characterizing (coupled) electro-thermal effects.

The characterization of the series resistances from dedicated sheet resistance structures is limited to the base and collector resistance, since structures for the emitter resistance cannot be fabricated in this process. The base resistance is further separated into a bias dependent internal and a bias independent external resistance.

The sheet resistance of the internal base (r_{sBi}) is extracted from tetrode transistors with a ring emitter ([RS91]), i.e. transistors with two separate base contacts. The layout is similar to what is shown in [SL07], i.e. it comprises a rectangular ring emitter and base contacts at the sides. The layout is visualized in Fig. 4.9(a). Tetrodes with $b_{\text{E}} = [0.13, 0.26, 0.39, 0.78]\,\mu\text{m}$ and $l_{\text{E}} = [2.66, 5.16, 10.16]\,\mu\text{m}$ were designed in every combination. The dimensions were chosen to be consistent with actual transistor dimensions (cf. Tab. 4.1). Only four pads are provided since Kelvin type probes are used[1]. Resistances from the metal interconnects are assumed to be small.

Various structures for the components of the external base ([SKAA13]) are available: length specific contact resistance r_{KB}, the sheet resistance of the silicide r_{sSil}, of the poly-silicon base over mono-silicon r_{sPM}, of the poly-silicon base over the oxide between base and collector r_{sPO} and of the base link r_{sSp} are to be distinguished. All structures are realized as three-terminal structures with five pads, allowing an extraction without any influence of interconnects. For each base resistance component, two structures with different contact length (and, thus, different length of the corresponding area) were designed.

The external collector resistance (R_{Cx}) is defined by the sheet resistance of the highly doped collector region r_{sBL}[2] and the contact resistance r_{KC}. Due to the mesa structure of the transistors no sinker exists and the contact is located directly on top of the highly doped collector. Therefore, a length specific collector resistance is extracted rather than an area specific value. Additionally, standard transistors

---

[1] Contacting a single pad with two standard DC probes is also possible.

[2] Although in this technology no classical buried layer exists, the naming convention was kept.

were modified to extract and verify the collector sheet resistance as described in [RKP$^+$07]. The structure was obtained by adding a *SalBlock*-layer between the collector and base contacts to avoid a short circuit due to the silicide at the fore-sides of the base poly. In the same place, also an *nSDBlock*-layer was added to avoid any n^+ implant at the fore-sides of the base. The layout is given in Fig. 4.9(b).

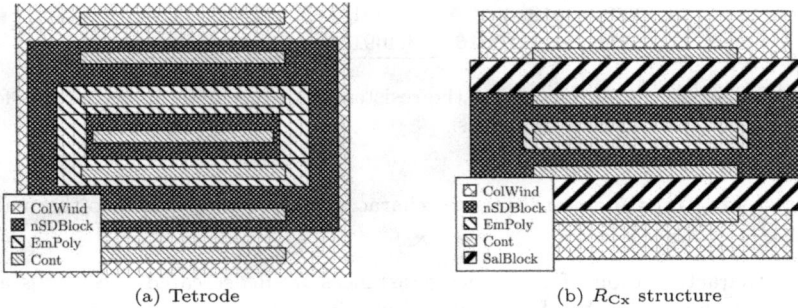

(a) Tetrode   (b) R_{Cx} structure

Figure 4.9: Layout of the tetrode for extracting r_{sBi} and the transistor for verifying R_{Cx}.

Thermal test structures are designed to allow the extraction methodology as described in [LZPS14]. Devices with four emitter fingers and different emitter lengths are available. They are not explained further here, since no extraction of the coupled thermal effects was performed.

4.3 Selected results from parameter extraction

In this section, only a brief overview of the parameter extraction for the process is given with a strong emphasis on the effects discussed in chapter 2 and 3. Since several of the experimental results from theses chapters were already presented for this process, they are not repeated here. A lot of publications exist for the extraction of specific parameters as well as for the complete parameter extraction of one or more processes. It is not the aim of this work to provide documentation of the extraction for the given process. Other works (e.g. [Kra15]) with a strong emphasis on parameter extraction methods are better suited for this purpose.

The aim of the model parameter extraction and device characterization of the presented process was to provide a model card for integration into the PDK. Therefore, only versions of HICUM/L2 which are available in commercial circuit simulators can be utilized. During the course of this work version 2.30 was widely available including several of the model equations from chapter 2.4. Yet, not included in this version[1] are the models for

- the $V_{B'C'}$ dependence of h_{f0},

- a model for NBR,

- the conductivity modulation of R_{su} as well as the extended substrate coupling network and

- a model for the electric field.

However, application of the $h_{f0}(V_{B'C'})$ was presented in 3.3.2 with data originating from this process. The application of the electric field model and the extended substrate network were also presented in the corresponding chapters for this process. The model the R_{su} will be briefly discussed in this section, although it is not included in the PDK model card.

Three different variations of the same generation of the process were available, because this test chip was part of the DOTSEVEN project [DOT14] including a strong emphasis on process development. The results presented in the following are for a process variation showing slightly improved performance. The device characterization was executed for all variations. However, non-linear measurements were only performed for the selected variation. In order to maintain consistency, only these results are presented

4.3.1 Overview

Presented in this section is a short discussion on the device scaling employing (3.21) and the scaling of the reverse Early effect. More details on the extraction of the scaling parameters are provided in [PSF13]. Although the results in this publication were obtained from the previous test chip, they are still almost the same for the

---

[1]These effects were also not available in the minor releases 2.31, 2.32 and 2.33 which were partially available in circuit simulators during the course of this work.

actual test chip. This was expected, because the previous test chip was designed for the same technology generation. Occurring variations did not significantly affect the scaling behavior.

The sheet resistances of the internal and external base and the collector were extracted using standard methods described in [RKP+07,RPZ+08] and from the TLM structures. The emitter and the thermal resistance were extracted employing the method described in 3.2. The obtained results are presented in this section. The transit time parameters of the transistors are extracted using [AZB+01,RSPL13] with a scaling of τ_0 performed as described in [SW96]. Although several geometry independent values such as built-in voltages of the capacitance components and non-ideality factors of the base current components were extracted from scaling components, they were fine-tuned for the reference device PDK_Nx_8. Additional independent values as for instance the weight factors and smoothing parameters were also extracted from this particular device and the CBEBC transistor with reference geometry. However, since best agreement should be obtained for PDK_Nx_8, results for the scaled CBEBC transistors are not always perfect (cf. 4.4.4).

4.3.2 Device scaling

The device scaling behavior is investigated based on the collector current. For a description of the subsequent usage of the scaling parameters for I_T and, thus, the effective emitter area see also [PSF13]. The results from the perimeter over area scaling of the CBEBC devices from Tab. 4.1 are shown in Fig. 4.10. Although almost ideal linear scaling is obtained for all devices (except for the marked devices with smallest dimensions), no physical values can be obtained due to the negative slope for width scaled devices and the negative intercept with the y-axis for length scaled devices.

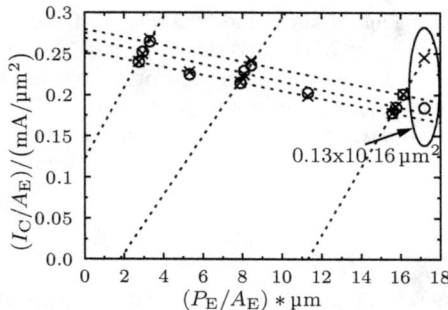

Figure 4.10: Scaling of the collector current for different devices using (3.20). Here, circle result from measurements and cross after application of (3.21). Dashed lines show the scaling for different $b_{E,0}$ and $l_{E,0}$. Results are shown for $V_{BE} = 0.6\,\mathrm{V}$.

In order to achieve a general scaling, (3.21) is applied to extract scaling parameters for the collector current. The results after extraction are also shown in Fig. 4.10, presenting a good agreement for all transistors. However, as expected from the P/A scaling, different values for $I'_{C,b}$ and $I'_{C,l}$ are obtained with $I'_{C,l} < 0$. Therefore, the method as demonstrated in 3.4.1.4 is also applied to these values. The obtained results are visualized in Fig. 4.11, showing that no solution is found for the given extracted data.

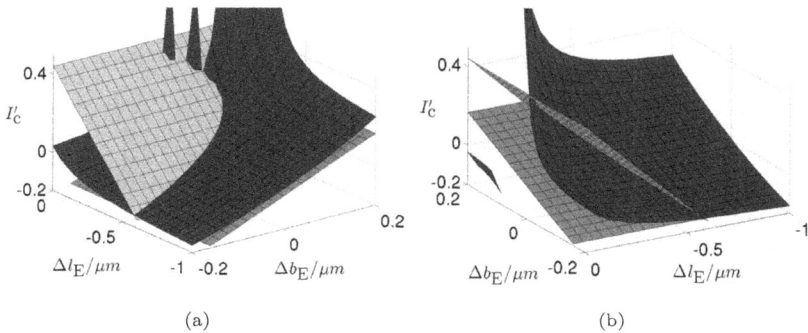

(a)   (b)

Figure 4.11: Visualization of (3.26) when solving for I'_C and Δb_E and Δl_E. Two different angles of view are shown here. The same operating point as in Fig. 4.10 is used.

An explanation of a possible cause for the observed issues is provided next, although it cannot be verified by actual pictures of the emitter window. Available pictures show emitter windows that are not presenting results after all finalizing process steps. The scaling was performed by defining the offsets of the emitter window.

The offsets are calculated according to (3.33), i.e. by $I'_{C,b}/\overline{I_{C,a}}$ and $I'_{C,l}/\overline{I_{C,a}}$. The extraction of the scaling parameters was repeated with the updated dimensions, obviously yielding $I'_{C,b,corr}$ and $I'_{C,l,corr}$ close to zero. The additional index "corr" is used to distinguish the new values from the original. $\overline{I_{C,a}}$ was not affected as mentioned in 3.4.1. Utilizing these offsets the P/A plot provides an almost horizontal line where deviations are caused by statistical variations, because both perimeter components are artificially forced to zero.

The value of $I_{C,c}/\overline{I_{C,a}}$ has the unit of an area. Due to the eliminated perimeter components, this area (labeled subsequently as $A_{E,c}$) can be interpreted as additional area that is not included in the assumed emitter dimensions. The bias dependence of this area is given in Fig. 4.12(a), showing an almost constant curve versus $V_{B'E'}$. This independence of the bias emphasizes the assumption of an additional area.

For the extracted numerical values, the following thought experiment can be performed. The extracted area is of approximately $0.02\,\mu m^2$, while the offset Δb_E is

30 nm. Since $I_{C,c}$ comprises effects from all four corners, the evaluated area is only a quarter. Similarly, only half of Δb_E is used, because the latter is calculated from $I'_{C,l}$ which is the sum of both sides. Assuming a rectangular shape of the additional areas, the remaining side is of approximately 333 nm.

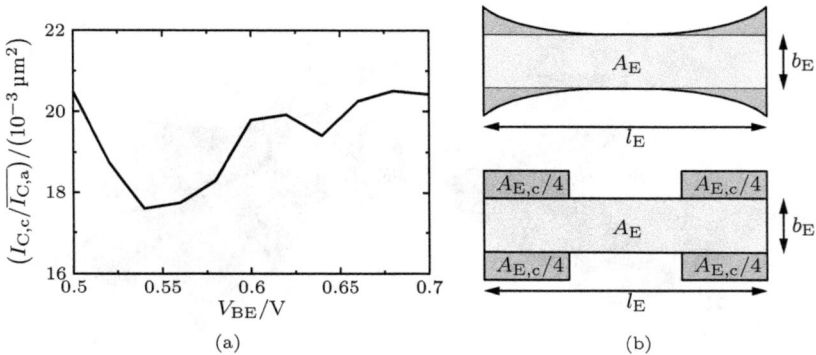

Figure 4.12: (a) Bias dependence of the additional area indicated by $I_{C,a}$. (b) Assumed emitter geometry.

Based on these results the assumed emitter geometry is sketched in Fig. 4.12(b). The top picture represents a realistic shape while the bottom picture shows the assumed simplified geometry. $A_{E,c}$ is caused by serifs near both fore-sides of the emitter window. They are created by a non-constant emitter window narrowing at the sides and lead to the negative $I'_{C,l}$ indicated by the negative slope in Fig. 4.10. This effect might also be an explanation for the non-ideal linear scaling for the smallest device, because it might occur varyingly strong for very short devices.

For creation of a scalable model, $b_{E,eff}$ and $l_{E,eff}$, both taking Δb_E and Δl_E from the original scaling into account, as well as $A_{E,c}$ are used to define the effective emitter area. For the smallest device an additional correction term is required.

4.3.3 Emitter and thermal resistances

The emitter and the thermal resistance are extracted employing the method introduced in chapter 3.2. R_E scales with the inverse emitter area where due to the unknown actual emitter geometry the drawn values are inserted. Results are presented in Fig. 4.13(a). There, ρ_E is the area specific emitter resistance. A comparison is shown for the extraction with and without utilizing previously extracted R_{Cx} values. For smaller devices the same results are obtained whereas small deviations occur for larger devices, especially with increasing b_E. The effect is caused by the scaling of R_E and R_{Cx}. With increasing b_E R_E decreases due to the large area while R_{Cx} slightly increases. Thus, the influence of R_{Cx} in (3.3) also increases with b_E. Note that the thermal model corresponding to $flsh=1$ was used. Except for small deviations for the largest device, the trend of R_E closely follows the anticipated behavior.

The scaling of R_{th} is based on the equation ([Sch00])

$$R_{th} = r_{th} f_{th}, \tag{4.1}$$

with the specific thermal resistance r_{th} and the geometry function

$$f_{th} = \frac{\ln (4l_E/b_E)}{l_E}. \tag{4.2}$$

Also for scaling the thermal resistance the drawn dimensions are inserted, yielding the results given in Fig. 4.13(b). An almost ideal scaling behavior is obtained for the extracted R_{th} values.

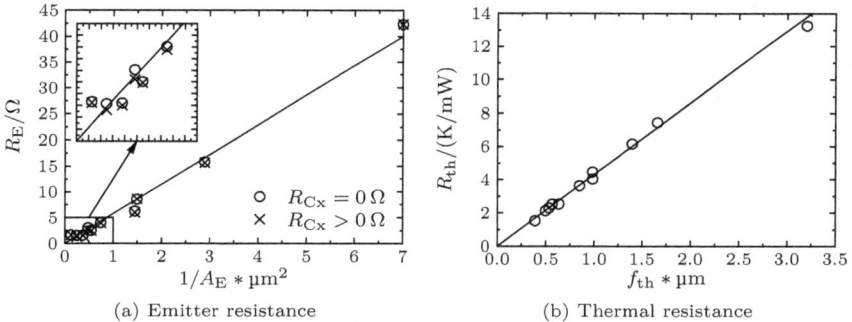

(a) Emitter resistance   (b) Thermal resistance

Figure 4.13: (a) Extraction results for R_E (circles and crosses). The solid line corresponds to an application of $R_E = \rho_E/A_E$. (b) Scaling of the extracted thermal resistance (circles) and application of (4.1) (solid line). Results are shown for CBEBC devices in both plots.

4.3.4 DC parameters

4.3.4.1 Terminal currents

The extraction of all terminal currents, i.e. I_C, I_B, I_E and I_S, was performed on transistors in RF-GSG pads. Although transistors in DC pads with separate contacts for all terminals were available, it was not possible to apply the following sweeps without damaging them even at medium bias. Also, since geometry scaled devices were only available in GSG-pads, it was required to extract the currents from those devices.

Performing different voltage sweeps allows to separate emitter and substrate current from known physical behavior. Measurements with compact model curves are given Fig. 4.14. Shown here are already the final compact models in order to emphasize effects that cannot be modeled by the actual version of HICUM/L2. As shown in Fig. 4.14(a) for negative V_{BE}, not only a tunneling current between base and emitter[1] but also between base and collector exists. At reverse bias, tunneling currents are dominated by band-to-band (BTB) tunneling. BTB tunneling is the most likely cause of this current due to the increased collector doping. This is also verified in Fig. 4.14(b), showing base and collector current at reverse bias superimposed upon each other and with the same numerical values as in Fig. 4.14(a). In the existing HICUM/L2 no model equations for this component are included. However, the same equations as for BE-BTB tunneling could be employed but the corresponding operating regions are of very little interest.

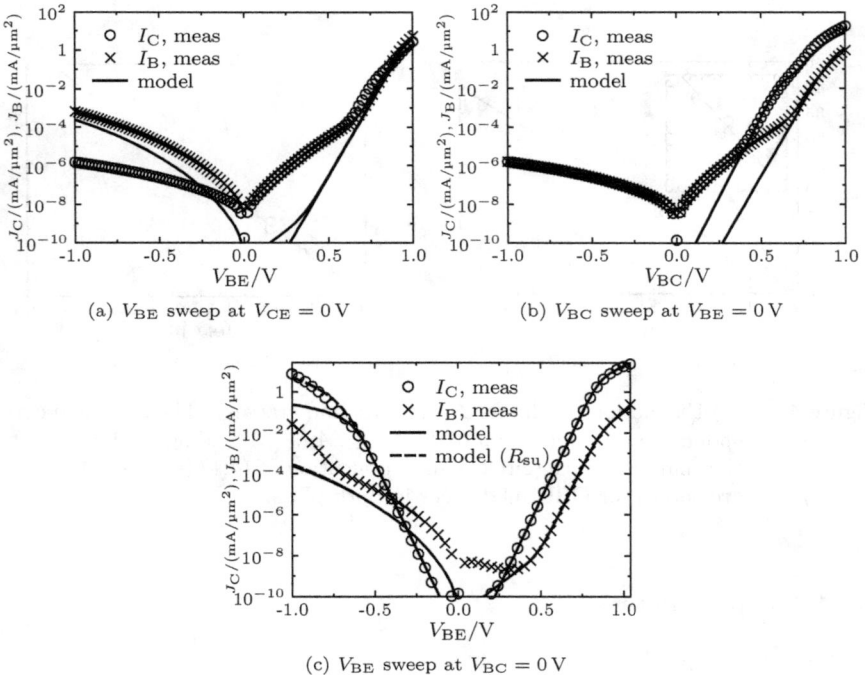

(a) V_{BE} sweep at $V_{CE} = 0\,\mathrm{V}$

(b) V_{BC} sweep at $V_{BE} = 0\,\mathrm{V}$

(c) V_{BE} sweep at $V_{BC} = 0\,\mathrm{V}$

Figure 4.14: Base and collector current for different voltage sweeps. In (c) "model R_{su}" includes (2.200) for demonstration purposes.

In Fig. 4.14(a) and (b) also a large non-ideal BC current at medium forward bias

---

[1] With the actual formulation of the HICUM/L2, no agreement for the BE tunneling current could be obtained, since the extracted parameters, more precise the saturation current, is outside of the valid range. Thus, it is no model but just an implementation issue. In future versions, the upper limit is most likely increased.

of V_{BC} is visible, caused by either tunneling or recombination effects. Also here, no compact model equations exist to describe this effect. However, in contrast to the BC BTB tunneling, corresponding operating regions affect standard operating regions, especially saturation. As shown later in 4.4.1, this region cannot be modeled accurately due to this effect. Standard forward operation is shown in Fig. 4.14(c). At low forward bias a non-ideal base current component exists. The same is also obtained for different SiGe technologies ([Der14]). From the bias dependence of this effect, most likely trap-assisted-tunneling (TAT) is causing this effect. Although several publications on TAT are available, most focus on MIS processes. To the authors knowledge, no suitable compact model for bipolar transistors exists. However, since located in the same bias range, formulations for the recombination current can be used.

At reverse bias, BE tunneling is dominating the base current. For large reverse bias, the effect explained in 2.6.4 causes the SC diode current as well as the (reverse) substrate transfer current to assumed small values. A demostration of the model (2.200) is inserted as dashed lines, showing the improved model for the substrate current. The voltage feedback of I_{SC} over R_{Cx} causes the transfer current to not exceed the tunneling current.

4.3.4.2 Transfer current

In contrast to other high-speed SiGe technologies, the SG13G2 technology shows the g_m reduction discussed in chapter 2.4.1.2 only to a very small extent. Due to the very small impact of a_{hjEi} its extraction is too error prone and was therefore omitted for this technology. According to that, in the scope of this section only h_{jEi} rather than h_{jEi0} is used for labeling the parameter[1].

The extracted $h_{jEi}(T)$ is given in Fig. 4.15(a). The extracted values follow the physics-based trend discussed in chapter 2.4.2.4. The model (2.139) is perfectly suited to model the decrease of h_{jEi} with increasing ambient temperature.

The method for scaling the weight factor h_{jEi} introduced in chapter 3.4.2 was applied to this technology. Using the previously extracted scaling parameters for I_T, the Early voltage was scaled by utilizing (3.46). Results are summarized for a b_E variation at a single l_E in Fig. 4.15(b), but similar results are obtained for the remaining geometries. The resulting value for h_{jEi} was calculated from V_{Er} and the area component of the zero-bias charge $\overline{Q_{p0a}}$ and the BE-depletion capacitance $\overline{C_{jEa0}}$.

The extraction results for the low current minority charge weight factor h_{f0} and the results for the weight factors for the high current minority charges h_{fE} and h_{fC} were already presented in chapter 3.3, more specifically in figures 3.11-3.14 and are not repeated here.

---

[1] This is also consistent with the compact model. In its formulation *hjei* was kept as parameter name to provide backward compatibility

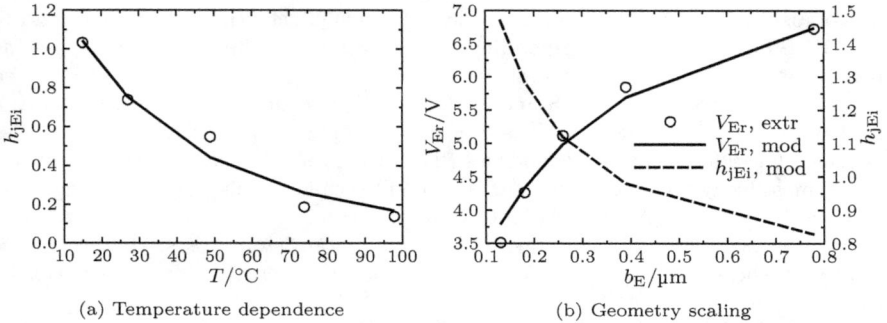

(a) Temperature dependence   (b) Geometry scaling

Figure 4.15: (a) Extraction results for h_{jEi} for different temperatures and the application of the model (2.139). (b) Scaling of V_{Er} and h_{jEi} versus b_E at $l_E = 10.16\,\mu m$.

4.4 Model results

In this section, a number of selected results for the extracted compact model are given. For a scalable technology the results of the compact model are a function of the static and dynamic operating bias point, the ambient temperature and the device geometry. Therefore, the number of possible verification plots can be very high. In this section, only the most relevant plots are presented. Additional important plots are presented at the end of the document in App. B.2.

The comparison provided in this work contains results from a single transistor (reference PDK device with $N_x = 8$) for a large bias[1] and temperature range. Also shown are selected results for the scaling of the CBEBC devices versus b_E and l_E. Finally, non-linear measurements with different measurement systems were performed. Compact model results in both frequency- and time-domain based on these data will be given.

4.4.1 DC characteristics

An overview of the model for the terminal currents was already provided in chapter 4.3.4.1. In the following, the focus is only on the most relevant active forward region. The modeling at low injection provides very accurate curves, although deviations exist at very low bias and positive V_{BC}. This is directly correlated to the non-ideal BC effects discussed in 4.3.4.1.

In Fig. 4.17 a more detailed plot of the currents at high injection is given. The modeled collector current shows a slightly larger spread with V_{BC} at very high input voltages. This can be caused by small inaccuracies of the temperature effects, either R_{th}, $h_{fC}(T)$ or $R_{Cx}(T)$. However, the overall agreement is very good and the

---

[1]All plots are given for room temperature if not noted differently.

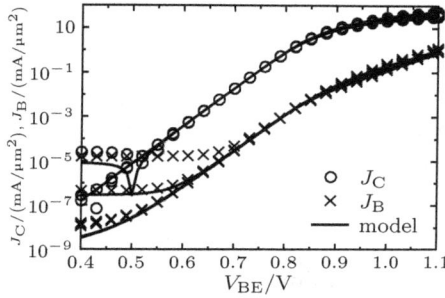

Figure 4.16: Base and collector current density for $V_{\text{BC}} = [-0.5 \ldots 0.5]\,$V.

relative error is less than 10% in the whole high current region. Larger deviations are obtained for the base current. However, also including the NBR model for high injection (parameter τ_{Bhrec}) the overall modeling capabilities of HICUM/L2 are by far not as elaborate as for I_{C}. This is justified by the fact that the base current at very high injection is less relevant, since the transistor is only rarely operated in this region. If anything only dynamic large signal operation may reach within this region. For all relevant cases the fundamental frequency of the circuit in conjunction with the strong charge increase leads to a domination of dynamic currents. Thus, no more effort was spent on optimizing the DC base current.

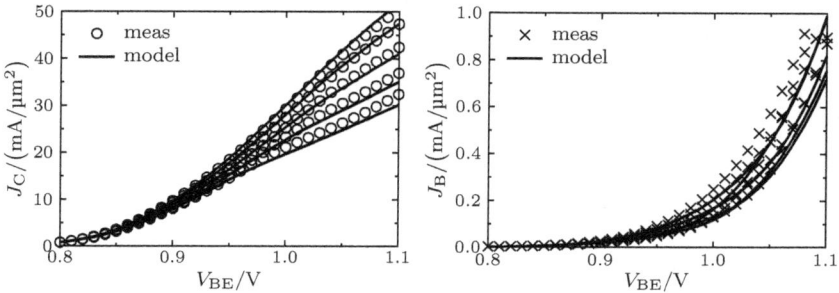

Figure 4.17: Zoom-in of the base and collector current density for a base-collector voltage range $V_{\text{BC}} = [-0.5 \ldots 0.5]\,$V.

4.4.2 Small-signal operation

Small-signal operation is verified based on small-signal Y-parameters as function of operating point and frequency. Furthermore, important FOMs of the transistor are derived and presented next. In the figures, the derived quantities are defined as follows. The terminal capacitances are given by

$$C_{\mathrm{BE}} = \frac{\Im\{\underline{Y}_{11} + \underline{Y}_{12}\}}{2\pi f} \quad \text{and} \quad C_{\mathrm{BC}} = -\frac{\Im\{\underline{Y}_{21}\}}{2\pi f}. \tag{4.3}$$

The cutoff frequency is given by

$$f_{\mathrm{T}} = \frac{f}{\Im\{\underline{Y}_{11}/\underline{Y}_{21}\}}. \tag{4.4}$$

Following from Masons unilateral power gain [Mas54]

$$U = \frac{|\underline{Y}_{21} - \underline{Y}_{12}|^2}{4(\Re\{\underline{Y}_{11}\}\Re\{\underline{Y}_{22}\} - \Re\{\underline{Y}_{12}\}\Re\{\underline{Y}_{21}\})}, \tag{4.5}$$

the maximum oscillation frequency

$$f_{\mathrm{max}} = f\sqrt{U} \tag{4.6}$$

is calculated. Although the latter two quantities are extracted in this work at a specific spot-frequency they can also be obtained from an extrapolation of the dynamic current gain $|\underline{h}_{21}|$ and U to 1, respectively.

4.4.2.1 Available measurements

Two measurements of S-parameters were performed. The standard setup comprises the Keysight PNA N8361C and Cascade |Z|-mx-probes, allowing measurements from 100 MHz to 67 GHz. The second setup uses Cascade Infinity I-110 probes and the N5260 millimeter-wave controllers for the PNA, extending the available frequency range to 110 GHz. The measurements for the parameter extraction were performed using the standard setup, while only a few verification measurements were performed with the increased frequency range. Therefore, plots for the derived quantities are always given for the standard measurements due to the higher number of available operating points.

While for the standard measurements only SOLT calibration was performed, measurements for the high frequencies are available with SOLT, LRRM and LRM+ calibration. A short comparison between the results is shown in B.1. Since for all three calibration methods almost the same results were obtained only results from the LRM+-calibration are used in the following plots.

In Fig. 4.18 both measurements are compared for a deembedding open and short. For the open a slightly less ideal behavior is obtained from the 110 GHz setup. $\underline{S}_{11}$ and $\underline{S}_{22}$ are closer to the border of the smith chart for the 67 GHz measurements. The effect is even stronger for the short where a significantly larger series resistance (shift on the real axis) is obtained. Based on the results from Tab. 4.2, the lower resistance for the short obtained by the 67 GHz measurements seem to be much more correct. Possible causes of the issues from the extended setup are not discussed here further.

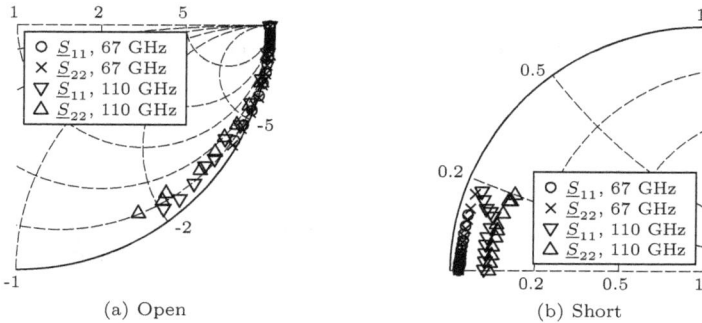

Figure 4.18: Measured S-parameters for the open and short utilized to deembed PDK_Nx_8. Comparison is shown here between 67 GHz and 110 GHz measurements.

4.4.2.2 Derived quantities

From the quantities defined in (4.3)-(4.6) only f_T and f_{max} are provided here. Both terminal capacitances are shown in the appendix. Results for the former are presented in Fig. 4.19. Only small deviations for the onset of the Kirk effect and, thus, the maximum of f_T and f_{max} for forward V_{BC} exist. Additionally, the fit of f_{max} for low currents has only a lower accuracy. However, for both a very good overall agreement is obtained.

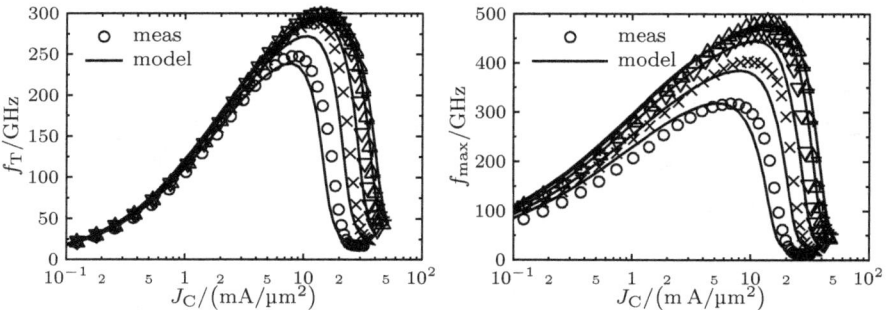

Figure 4.19: Cutoff frequency calculated from (4.4) at a spot frequency $f = 10\,\text{GHz}$ and f_{max} from (4.6) at 20 GHz. Both are shown in the V_{BC} range of Fig. 4.16.

4.4.2.3 Small-signal parameters

Verification plots for all Y-parameters as a function of collector current and frequency are provided in B.2.2.2. The comparison is carried out showing results from both measurement setups. In order to compare the setups, similar operating points and frequencies are used. However, as for the previous discussion of the deembedding structures, also for the actual device different results for the same operating regions were obtained, with more reasonable results from the 67 GHz-system.

A very good overall agreement for the results from the 67 GHz-system is obtained. Errors are still present for the modeling of $\underline{Y}_{22}$ due to the effects discussed in 2.6. No further effort was spent on optimizing the curves for the 110 GHz-system due to the questionable results.

More accurate measurements up to 110 GHz and additional measurements in the higher frequency bands are required for a more comprehensive verification of the high-frequency modeling. However, they were not available at the time of this work. Therefore, a further discussion is not meaningful in this place.

4.4.2.4 Dynamic emitter current crowding

Only a very brief discussion of the dynamic emitter current crowding is provided. The device with $b_E = 0.78 \, \mu m$ and $l_E = 5.16 \, \mu m$ is used, since the lateral NQS effect has negligible effect on the smaller devices. For this frequency range, employing $f_{CRBi} > 0$, i.e. including the lateral NQS effect, significantly improves the shape of the curves. However, for a more elaborate discussion measurements at very high frequencies (upto 110 GHz and beyond) are required.

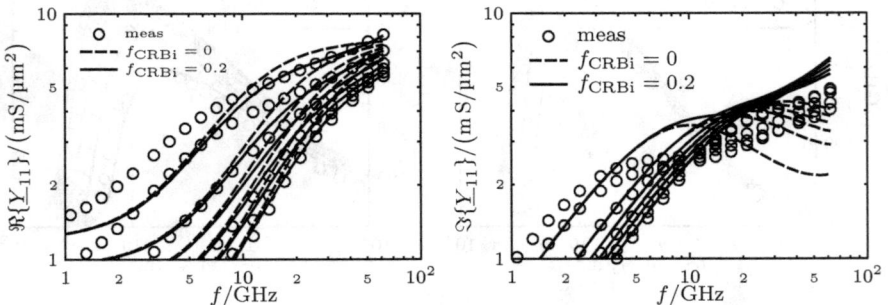

Figure 4.20: Real and imaginary part of $\underline{Y}_{11}$ of `1x_CBEBC_0p78x5p16` and $V_{BE} = [0.9 \ldots 1.0]$ V, showing the impact of the model of the lateral NQS effect.

4.4.3 Temperature dependence

The temperature dependences of the base and collector current as well as of f_T, which are presented in Fig. 4.21, are not given for the actual process version. Temperature parameters were assumed to be independent of the process version. Therefore and due to limited time slots at the measurement equipment small-signal measurements were only performed for the reference version that is slightly slower. Although for verification purposes DC characteristics were measured for all variations, consistent results for one transistor are given here. The DC curves show a very good modeling for the given temperature region.

Additionally, due to restrictions in the possible setups the measurements were performed employing the HP-8510C network analyzer which often gives much less accurate results in terms of measurement noise. Therefore, the measured f_T curves presented in the figure are not giving clear trends in terms of an almost linear dependence on the temperature. Due to the same reason no f_{max} curves are provided for different ambient temperatures. Obtained reference results are far too error prone. One final note for Fig. 4.21 is that the underlying model card was fine-tuned for the device at a different location of the same die. Due to worn.out pads, one of the backup devices was measured showing slightly different f_T values.

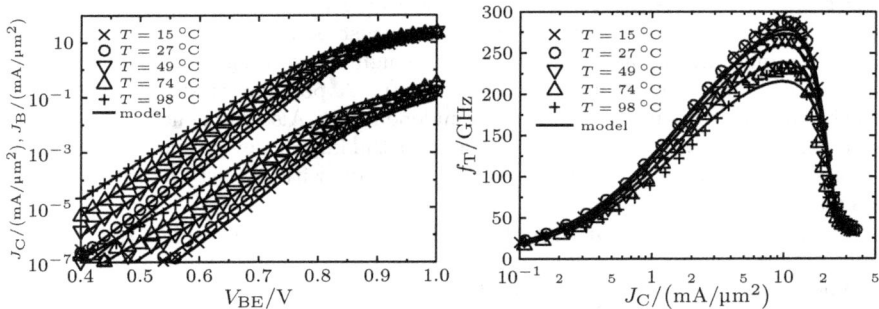

Figure 4.21: Forward gummel plot and f_T for PDK_Nx_8 of the reference process variation versus ambient temperature. Both curves are plotted for $V_{BC} = 0\,V$.

4.4.4 Scaling

Verification of the scaling equations utilized for the model card of the CBEBC transistors is given for a b_E variation at $l_E = 10.16\,\mu m$ as well as for an l_E variation at $b_E = 0.13\,\mu m$. For the corresponding swept values see Tab. 4.1, presented results are the collector current and transit frequency. In appendix B.2.3 additionally the base current, f_{max} and both terminal capacitances are provided.

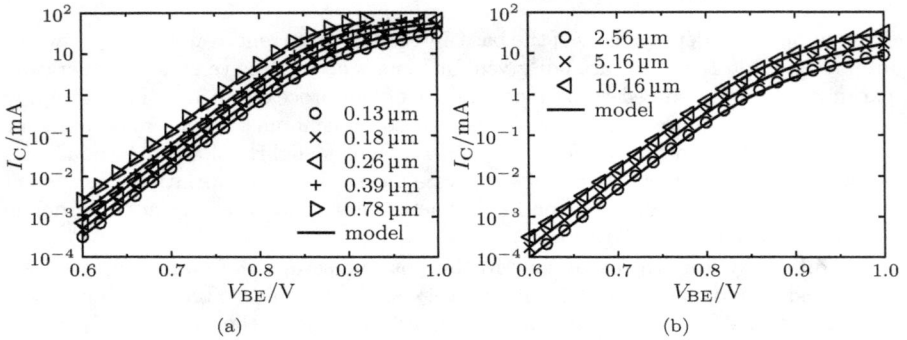

Figure 4.22: Collector current at $V_{BC} = 0\,V$ and room temperature for (a) a scaling versus emitter width and (b) emitter length.

A very accurate I_C scaling versus both b_E and l_E is obtained, verifying the scaling rules introduced in 3.4 and applied in 4.3.4.2. The results for f_T versus l_E also show a good agreement. Generally, for geometry variations with maintained $l_E \gg b_E$ scaling versus emitter length almost always provides accurate results, since the P_E/A_E ratio only slightly changes. In contrast, scaling of f_T versus b_E shows increasing inaccuracies at medium bias although the scaling of the capacitances provides good fits (cf. figures B.11 and B.12). A possible reason is the emitter transit time which is modeled geometry independent. Also the general trend of f_T as increasing with b_E and finally decreasing for the largest device could be modeled, but not as accurately as for the previous test chip with almost the same process version shown in [PSF13][1].

The additional results in App. B.2.3 show a very good scaling of I_B, C_{BE} and C_{BC}, although with small deviations in the bias dependence. These are caused by the fine-tuning to the reference (non-CBEBC) device. The scaling of f_{max} shows almost constant values versus l_E due to the reasons discussed before. The results for the b_E-scaling are not as accurate, which can also be caused by inaccuracies during measurements rather than by model issues. The uncertain trend of the measured values proves this assumption. Hence, no further effort was spent on improving the results

4.4.5 Large-signal modeling

4.4.5.1 Available measurements

The device `PDK_Nx_8` was measured with non-linear equipment. Two sets of data are available, obtained by employing the Agilent PNA-x non-linear network analyzer

---

[1]However, the reference variation for the actual test chip shows the same behavior as given in Fig. 4.23.

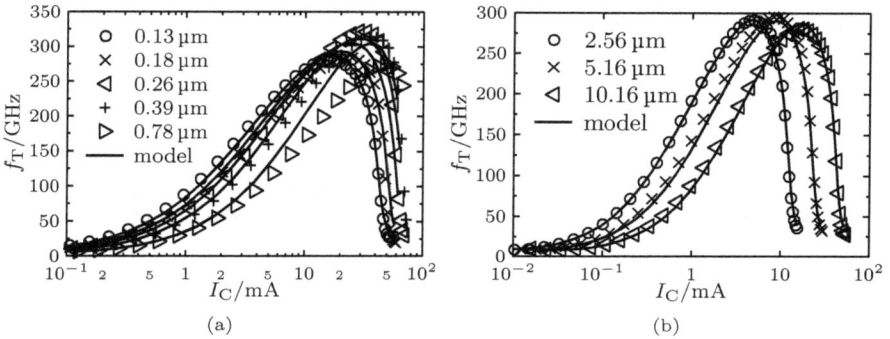

Figure 4.23: Transit frequency at $V_{BC} = 0\,V$ and room temperature for (a) a scaling versus emitter width and (b) emitter length at spot-frequency $f = 10\,GHz$.

and the linear Rhode & Schwarz ZVA-67 network analyzer, extended with non-linear active load-pull capabilities by adding a digital tuner from Mesuro [WWS+14]. A detailed discussion on the results from the PNA-x system follows subsequently.

Measurements were taken at fundamental frequencies of $f_0 = [2.5, 5.0, 10.0]\,GHz$ for a large set of operating points, ranging from $V_{BE} = 0.75\,V$ to $V_{BE} = 0.9\,V$ at $V_{CE} = 1.0\,V$. For the following discussion $V_{BE} = 0.875\,V$ is used which is close to the theoretical studies in 2.3.3. The operating point also being close to peak f_T is visualized in Fig. 4.24.

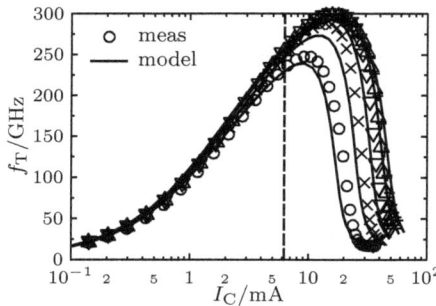

Figure 4.24: Operating point for non-linear modeling in the f_T curves, represented by the vertical dashed line. The model is shown for a verification of the quasi-static modeling.

However, the actual operating point in the circuit simulation is based on the collector current at a low input power. The base-emitter voltage is adjusted accordingly by 1-2 mV. This fine-tuning of the setup allows to eliminate small deviations due to process and temperature variations.

4.4.5.2 Steady-state simulations

A common way to simulate the given kind of circuit is by using steady-state simulations either in frequency domain with the Harmonic Balance approach (HB, [NV76,BWL95]) or the shooting method ([KWSV90]). In this work, HB simulations were employed, because the results are in a similar format as the experimental data.

In B.3.1 a short discussion on the number of required harmonics for calculating time-domain curves is presented. It is important to note from the results there that generally a larger number of harmonics should be included in the simulation setup. In the following, simulations are carried out employing 11 harmonic frequencies.

4.4.5.3 Source and load impedance

In contrast to simulations of circuits as in 2.3.3, the environment during measurements is not exactly known. Although the source network analyzer is in a $50\,\Omega$ environment, for example uncertainties during the non-linear calibration process can introduce errors in terms of load and source impedances.

Ideal elements in the bias tee are assumed, i.e. blocking capacitor and feed inductance are large enough. During circuit simulations power sources are represented by a voltage source and an internal impedance. The calculation of the source impedance $\underline{Z}_S$ from known P_{avs} and $\underline{V}_{BE}$ can be performed based on the circuit given in Fig. 2.19. The maximum available power is translated into $V_S = 2\sqrt{2P_{avs}\Re\{\underline{Z}_s\}}$. Thus, the equation

$$\frac{\underline{Z}_S + \underline{Z}_{bb}}{\underline{Z}_{bb}}\underline{V}_{BE} = 2\sqrt{2P_{avs}\Re\{\underline{Z}_S\}} \tag{4.7}$$

is solved with respect to $\underline{Z}_S$. In this equation

$$\underline{Z}_{bb} = \frac{\underline{V}_{BE}}{\underline{I}_B} \tag{4.8}$$

is the input impedance of the device. The equation is solved with respect to real and imaginary part, leading to the non-linear system of equations

$$\Re\{\underline{Z}_S\}\Re\{\underline{I}_B\} - \Im\{\underline{Z}_S\}\Im\{\underline{I}_B\} + \Re\{\underline{V}_{BE}\} - 2\sqrt{2P_{avs}\Re\{\underline{Z}_s\}} = 0$$
$$\Re\{\underline{Z}_S\}\Im\{\underline{I}_B\} + \Im\{\underline{Z}_S\}\Re\{\underline{I}_B\} + \Im\{\underline{V}_{BE}\} = 0, \tag{4.9}$$

which is solved with respect to the solution variables $\Re\{\underline{Z}_S\}$ and $\Im\{\underline{Z}_S\}$.

However, since P_{avs} only represents the generated power at the fundamental frequency f_0, (4.7) can only be applied for f_0. For all harmonic frequencies, the input impedance is calculated by

$$\underline{Z}_S = -\frac{\underline{V}_{BE}}{\underline{I}_B},\tag{4.10}$$

because the harmonic components of the base current cause a voltage drop across $\underline{Z}_S$, resulting in the harmonic components of $\underline{V}_{BE}$.

Based on the same assumptions, the load impedance Z_L is calculated as

$$\underline{Z}_L = -\frac{\underline{V}_{CE}}{\underline{I}_C}\tag{4.11}$$

for all frequencies. Results from actual measurements are given in Fig. 4.25 and 4.26.

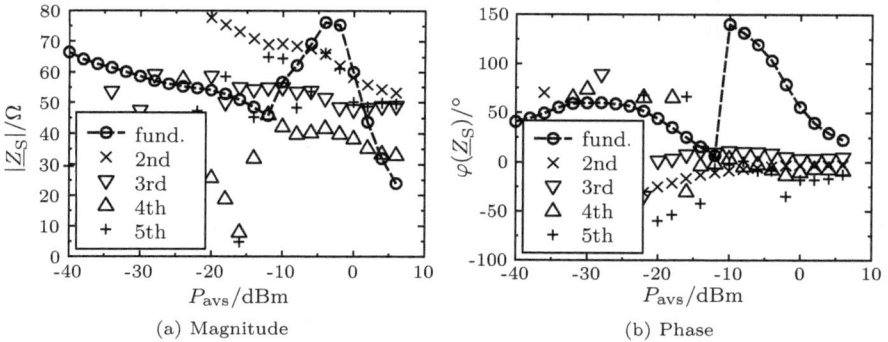

(a) Magnitude   (b) Phase

Figure 4.25: Magnitude and phase of the source impedance for the first five harmonic frequencies calculated using (4.7) and (4.10).

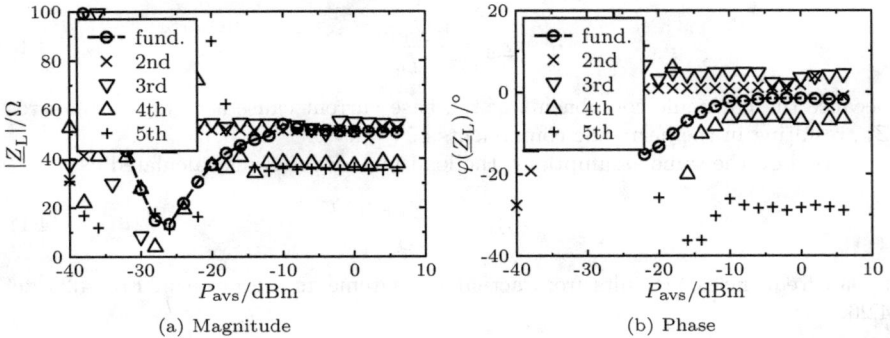

(a) Magnitude   (b) Phase

Figure 4.26: Magnitude and phase of the load impedance for the first five harmonic
frequencies calculated using (4.11).

4.4.5.4 Verification with small-signal parameters

Results obtained from non-linear measurements for lower input power can be verified
employing linear S-parameter measurements, more specifically using any two-port
representation. In this work Y-parameters are used. It is sufficient to neglect the
power source but use the measured fundamental component of V_{BE} directly as input for the circuit. The corresponding equivalent circuit is given in Fig. 4.27. It
directly incorporates Y-parameters, allowing a verification of the non-linear results
with measured Y-parameters.

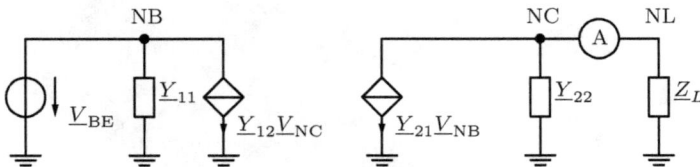

Figure 4.27: Equivalent circuit used for verifying the non-linear measurements with
linear Y-parameters.

The fundamental components from $\underline{I}_B$ and $\underline{I}_C$ as well as $\underline{V}_{CE}$ are calculated
using the system of equations derived with the MNA approach for above circuit.

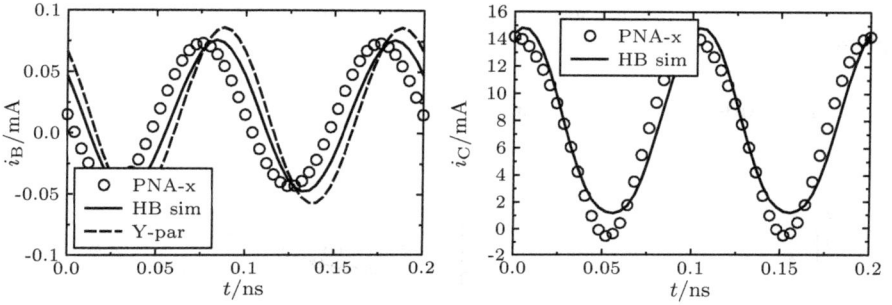

Figure 4.28: Time-domain total base current at $P_{\mathrm{avs}} = -35\,\mathrm{dBm}$ and total collector current at $P_{\mathrm{avs}} = -15\,\mathrm{dBm}$ for the fundamental frequency $f_0 = 10\,\mathrm{GHz}$.

The equations are

$$
\begin{bmatrix}
\underline{Y}_{11} & \underline{Y}_{12} & 0 & -1 & 0 \\
\underline{Y}_{21} & \underline{Y}_{22} & 0 & 0 & -1 \\
0 & 0 & 1/\underline{Z}_{\mathrm{L}} & 0 & 1 \\
1 & 0 & 0 & 0 & 0 \\
0 & 1 & -1 & 0 & 0
\end{bmatrix}
\begin{bmatrix}
\underline{V}_{\mathrm{NB}} \\
\underline{V}_{\mathrm{NC}} \\
\underline{V}_{\mathrm{NL}} \\
\underline{I}_{\mathrm{B}} \\
\underline{I}_{\mathrm{C}}
\end{bmatrix}
=
\begin{bmatrix}
0 \\
0 \\
0 \\
\underline{V}_{\mathrm{BE}} \\
0
\end{bmatrix},
\tag{4.12}
$$

with $\underline{V}_{\mathrm{CE}} = \underline{V}_{\mathrm{NC}}$. A comparison to the actual measurements at $f_0 = 10\,\mathrm{GHz}$ is provided in Fig. 4.28. The base current is given here for a very small input power and, thus, almost ideal linear behavior. However, a large phase shift is visible between the measured i_{B} (circles) and i_{B} calculated from measured Y-parameters and from v_{BE} (dashed lines)[1]. Such a phase-shift leads to large issues if not consistent for all harmonic components. Also shown in Fig. 4.28 is the time-domain total i_{C} exhibiting negative values. However, as theoretically discussed in 2.3.3 for a transistor of comparable f_{T} dynamic currents, which can be the only cause of $i_{\mathrm{C}}(t) < 0$, cannot occur at such small fundamental frequencies.

Following from the discussions in 2.3.3 10 GHz can be considered as quasi-static operation. In the following sections only results for 5 GHz are presented, because operation at this frequency can be considered as quasi-static as well and they do not inhibit measurement issues as the 10 GHz measurements. Selected additional results for 2.5 GHz and 10 GHz are provided in B.3.2. It is very unlikely that different physical effects are obtained, at least for the fundamental components. Furthermore, the highest measured component with 50 GHz is still far away from the cutoff-frequency at the given operating point.

Due to the inaccuracies of the phase for high frequencies the data from Mesuro

---

[1]The small deviations between non-linear (solid line) and the linear (dashed line) are caused by small errors in v_{BE} due to the non-constant $\underline{Z}_{\mathrm{S}}$. In simulations a fixed input impedance was used.

measurements are not presented. Although for these data the phase shift seems to be consistent for all harmonic components, even the input voltage suffers from unknown phase-shifts making the simulation setup too error prone to draw any conclusions.

4.4.5.5 Frequency-domain results

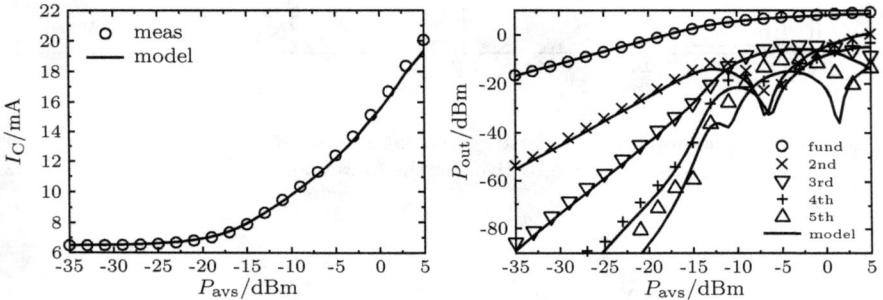

Figure 4.29: Operating point and fundamental and harmonic components of the output power as a function of available input power for the fundamental frequency $f_0 = 5\,\text{GHz}$.

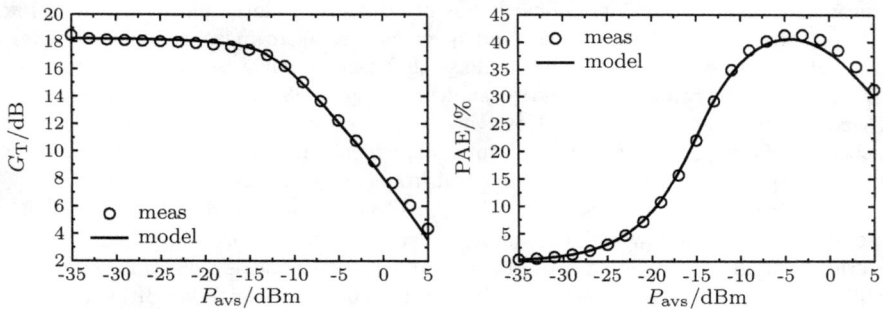

Figure 4.30: Transducer gain and PAE as a function of available input power for the fundamental frequency $f_0 = 5\,\text{GHz}$.

The comparison is carried out based on the quantities defined in 2.3.3.4. Figures 4.29 and 4.30 show important FOMs as a function of available source power. Very good agreement is obtained for all quantities in the complete power range. Small differences in the fourth and fifth harmonic component of the output power at high

input power are present. The accurate modeling of the fundamental output power
and operating point directly leads to accurate results for the power gain and PAE.

4.4.5.6 Time-domain results

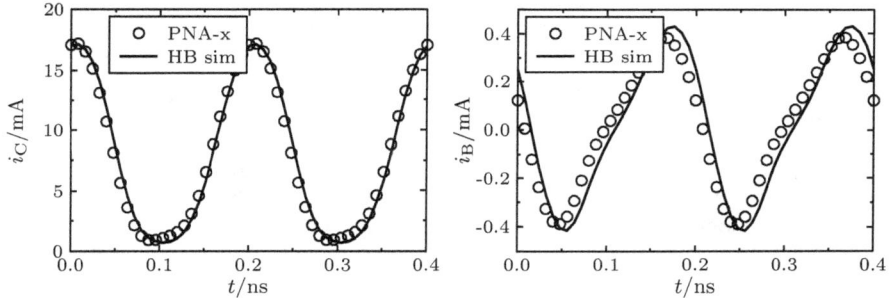

Figure 4.31: Time-domain collector and base current at $P_{\mathrm{avs}} = -15\,\mathrm{dBm}$ for the
fundamental frequency $f_0 = 5\,\mathrm{GHz}$.

The obtained time-domain base and collector currents are given for two input
powers in figures 4.31 and 4.32. The smaller input power corresponds to $P_{\mathrm{in,1dB}}$. At
this power a small phase shift is obtained for both curves. In figures showing $i_{\mathrm{B}}(t)$
values obtained from measured Y-parameters are inserted as dashed lines. As visible
by the falling edge the phase shift is also present for the linear base current at this
input power. It is therefore (and due to the relative small fundamental frequency)
considered as deviation of the measurement.

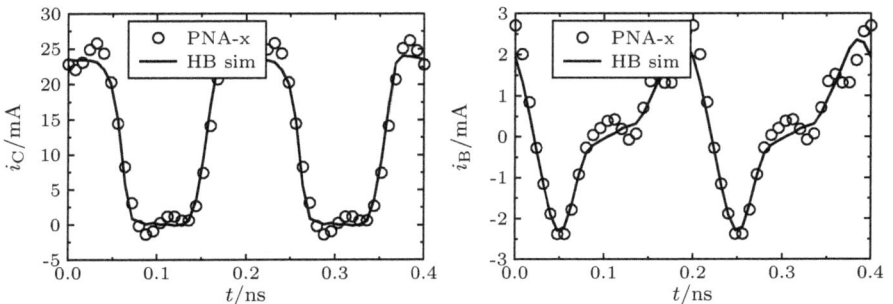

Figure 4.32: Time-domain collector and base current at $P_{\mathrm{avs}} = -5\,\mathrm{dBm}$ for the fundamental frequency $f_0 = 5\,\mathrm{GHz}$.

For the larger input power very accurate results are obtained for i_C and i_B. Note that the higher number of harmonics in HB simulations leads to smoother curves. However, due to the relative small frequency, the impact of NQS effects as discussed in 2.3.3.3 could not be evaluated for actual transistors.

4.4.6 Summary

The application of the latest public available HICUM/L2 version to very advanced SiGe-HBTs was presented in this section. For all measured curves very good agreements were obtained. Due to the smaller non-ideal g_m degradation of this process in comparison to other actual SiGe technologies it is less suitable to emphasize the corresponding model improvements. However, the model extensions were verified based on several technologies and the results shown in several publications (e.g. [PSK+11,PSF13,PS14a]). Furthermore, linear and non-linear measurements suffer from either a very limited frequency range or issues at high frequencies. Therefore, the non-linear operation at high frequencies could not be verified based on actual measurements.

5 | Summary and Outlook

In this work several aspects of SiGe-HBT compact modeling were discussed. It comprises the derivation of both physics-based compact model equations suitable for the usage in production type compact models and basic modeling approaches for specific transistor effects.

The main contribution of this work is based on the quasi-static transfer current modeling. Although the emphasis was laid on high-speed devices and effects at high frequencies, existing quasi-static formulations have suffered from their origin based on silicon transistors. Effects introduced by Germanium in the base are leading to both non-ideal bias and temperature dependences and, thus, are affecting the transfer current in all relevant operating regions. The physical origin of these effects was discussed in detail. Based on the GICCR model equations for the formerly bias and temperature independent weight factors were derived and successfully applied to experimental data. Due to the impact on all operating regions employing these equations is essential for accurate circuit design utilizing advanced SiGe-HBTs. Most of the model equations are derived and verified based on high-speed devices. However, their application is not limited to high-speed transistors.

A general modeling approach for the electric field in the collector based on transit times is presented. The focus on transit times provides a direct link to terminal quantities, as the field value itself cannot be measured directly. The operating regions mainly affecting the electric field, namely the punch-through case and the partially depleted collector, were discussed separately. Models for both regions were subsequently combined into a single model by introducing an additional model for the current and transit time defining the transition region. Accurate results for pure silicon transistors and partially accurate results for a SiGe-HBT were obtained, both based on numerical simulations. The feasibility of an application to experimental data was shown, but yielded only partial agreements. Several aspects of the model have to be addressed in the future. Among the most important are the yet missing evaluation of non-local effects and temperature dependences.

Based on a simple yet physical approach, the existing equivalent circuit from HICUM/L2 for substrate coupling effects was extended to address the impact of the isolation oxide. This network together with the presented scaling equations improves the modeling of the dynamic output conductance. However, the underlying 2D device simulations are capable of accurately describing substrate coupling only in a limited range. Especially the scaling equations may only be of limited use for actual 3D structures with more complicated substrate geometry. However, by the application of low frequency measurements to transistors and dedicated test structures scaling equations can be verified or improved in future.

Several enhancements to existing compact model equations were highlighted by the improved model results for existing high-speed SiGe-technologies. Nevertheless, this work also revealed topics for future investigations. Most importantly, revised models for both the vertical and the lateral NQS effects are required. Both suffer from only a limited applicability and disadvantageous implementation. However, the latter point is often not an issue of the model itself but of available framework in commercial simulators.

Improved model equations are useless without dedicated extraction methods for the newly introduced parameters. As a consequence, the second part of this work focused on the parameter extraction methodology. The continuous work on the weight factor models has led to methods which enable the physics-based extraction of the transfer current parameters. A widely used existing approach was extended with an emphasis on the newly integrated physical effects. Using this improved method allows to extract all relevant parameters at both low and high injection explicitly taking thermal effects into account. The method was applied to a large set of latest generation SiGe-HBTs providing results which prove both the physics-based approach of the derived model equations and the applicability of the method. The contributions to geometry scaling laws of the transfer current provide a tool for a simple yet physics-based generation of scalable model cards.

A chapter of this work was dedicated to the emitter resistance extraction. In contrast to most existing methods where thermal effects are limiting the possible accuracy, self-heating was explicitly taken into account by the method presented this work. This approach is perfectly suited for current high-speed SiGe-HBTs which suffer from extreme thermal effects. By utilization of thermal effects the effective thermal resistance of the HBT is extracted as well. The usage of standard test structures and measurement setups allows the straightfoward integration of the method into existing extraction flows. As for the transfer current extraction, this method was also applied to several advanced SiGe-HBTs.

Finally, the derived model equations were applied to a selected high-speed SiGe HBT process developed by IHP. Highly accurate models for DC- and small-signal as well as for large-signal characteristics are presented and prove that HICUM/L2 including the new model formulations is perfectly suited to model very advanced SiGe-HBTs. Hence, this work significantly improves the state-of-the-art in SiGe HBT compact modeling.

Bibliography

[ARS+10]   ARDOUIN, B. ; RAYA, C. ; SCHROTER, M. ; PAWLAK, A. ; CELI, D. ;
           POURCHON, F. ; AUFINGER, K. ; MEISTER, T.F. ; ZIMMER, T.: Mod-
           eling and parameter extraction of SiGe:C HBTs with HICUM for the
           emerging terahertz era. In: *European Microwave Integrated Circuits
           Conference (EuMIC)*, 2010, pp. 25–28

[AZB+01]   ARDOUIN, B. ; ZIMMER, T. ; BERGER, D. ; CELI, D. ; MNIF, H. ;
           BURDEAU, T. ; FOUILLAT, P.: Transit time parameter extraction for
           the HICUM bipolar compact model. In: *Bipolar/BiCMOS Circuits
           and Technology Meeting (BCTM)*, 2001, pp. 106–109

[BCS+02]   BERGER, D. ; CELI, D. ; SCHROTER, M. ; MALORNY, M. ; ZIMMER, T. ;
           ARDOUIN, B.: HICUM parameter extraction methodology for a single
           transistor geometry. In: *Bipolar/BiCMOS Circuits and Technology
           Meeting (BCTM)*, 2002, pp. 116–119. – ISSN 1088–9299

[BSA+12]   BALTEANU, A. ; SARKAS, I. ; ADINOLFI, V. ; DACQUAY, E. ; TOMKINS,
           A. ; CELI, D. ; CHEVALIER, P. ; VOINIGESCU, S.P.: Characterization of
           a 400-GHz SiGe HBT technology for low-power D-Band transceiver
           applications. In: *IEEE International Microwave Symposium Digest
           (MTT)*, 2012, pp. 1–3. – ISSN 0149–645X

[BWL95]   BRACHTENDORF, H.G. ; WELSCH, G. ; LAUR, R.: Fast simulation
           of the steady-state of circuits by the harmonic balance technique.
           In: *IEEE International Symposium on Circuits and Systems (ISCAS)*
           vol. 2, 1995, pp. 1388–1391

[CB91]   CHO, H. ; BURK, D.E.: A three-step method for the de-embedding of
           high-frequency S-parameter measurements. In: *IEEE Transactions on
           Electron Devices* 38 (1991), no. 6, pp. 1371–1375. – ISSN 0018–9383

[CCLA+13] CAMILLO-CASTILLO, R.A. ; LIU, Q.Z. ; ADKISSON, J.W. ; KHATER,
           M.H. ; GRAY, P.B. ; JAIN, V. ; LEIDY, R.K. ; PEKARIK, J.J. ; GAM-
           BINO, J.P. ; ZETTERLUND, B. ; WILLETS, C. ; PARRISH, C. ; ENGEL-
           MANN, S.U. ; PYZYNA, A.M. ; CHENG, P. ; HARAME, D.L.: SiGe

HBTs in 90nm BiCMOS technology demonstrating 300GHz/420GHz f_T/f_{MAX} through reduced R_b and C_{cb} parasitics. In: *IEEE Bipolar/BiCMOS Circuits and Technology Meeting (BCTM)*, 2013, pp. 227–230. – ISSN 1088–9299

[CCP+93]   CRABBE, E.F. ; CRESSLER, J.D. ; PATTON, Gary L. ; STORK, J. M C. ; COMFORT, J.H. ; SUN, J. Y C.: Current gain rolloff in graded-base SiGe heterojunction bipolar transistors. In: *IEEE Electron Device Letters* 14 (1993), no. 4, pp. 193–195. – ISSN 0741–3106

[Cel09]   CELI, D.: An Attempt to Determine the Emitter Size of Bipolar Transistors from Electrical measurements. In: *9th HICUM Workshop, Würzburg, Germany*, 2009

[CMH+11]   CHEVALIER, P. ; MEISTER, T.F. ; HEINEMANN, B. ; VAN HUYLEN-BROECK, S. ; LIEBL, W. ; FOX, A. ; SIBAJA-HERNANDEZ, A. ; CHANTRE, A.: Towards THz SiGe HBTs. In: *IEEE Bipolar/BiCMOS Circuits and Technology Meeting (BCTM)*, 2011, pp. 57–65. – ISSN 1088–9299

[CPL+09]   CHEVALIER, P. ; POURCHON, F. ; LACAVE, T. ; AVENIER, G. ; CAMPIDELLI, Y. ; DEPOYAN, L. ; TROILLARD, G. ; BUCZKO, M. ; GLORIA, D. ; CELI, D. ; GAQUIERE, C. ; CHANTRE, A.: A conventional double-polysilicon FSA-SEG Si/SiGe:C HBT reaching 400 GHz f_{MAX}. In: *IEEE Bipolar/BiCMOS Circuits and Technology Meeting (BCTM)*, 2009, pp. 1–4. – ISSN 1088–9299

[CPS+90]   CRABBE, E.F. ; PATTON, G.L. ; STORK, J. M C. ; COMFORT, J.H. ; MEYERSON, B.S. ; SUN, J. Y C.: Low temperature operation of Si and SiGe bipolar transistors. In: *Technical Digest of International Electron Devices Meeting (IEDM)*, 1990, pp. 17–20. – ISSN 0163–1918

[Cre06]   CRESSLER, John D. (Hrsg.): *Silicon Heterostructure Handbook.* Taylor & Francis Group, LLC, 2006. – ISBN 978–0–8493–3559–4

[CT67]   CAUGHEY, D.M. ; THOMAS, R.E.: Carrier mobilities in silicon empirically related to doping and field. In: *Proceedings of the IEEE* 55 (1967), no. 12, pp. 2192–2193. – ISSN 0018–9219

[Der14]   DERRIER, N.: Experience of BE Tunneling current and BC Barrier effect with HICUM L2 v2.32. In: *Bipolar Arbeitskreis, Crolles, France*, 2014

[DOT11]   DOTFIVE: TOWARDS 0.5 TERAHERTZ SILICON GERMANIUM HETEROJUNCTION BIPOLAR TECHNOLOGY. *Homepage.* (http://dotfive.eu). 2011

[DOT14]   DOTSEVEN: TOWARDS 0.7 TERAHERTZ SILICON GERMANIUM
             HETEROJUNCTION BIPOLAR TECHNOLOGY. *Homepage*. (http://
             dotseven.eu). 2014

[DVHH+09a]   DECOUTERE, S. ; VAN HUYLENBROECK, S. ; HEINEMANN, B. ; FOX,
             A. ; CHEVALIER, P. ; CHANTRE, A. ; MEISTER, T. ; AUFINGER, K. ;
             SCHROTER, M.: Pushing the speed limits of SiGe:C HBTs up to 0.5
             Terahertz. In: *IEEE Custom Integrated Circuits Conference (CICC)*,
             2009, pp. 347–354

[DVHH+09b]   DECOUTERE, S. ; VAN HUYLENBROECK, S. ; HEINEMANN, B. ; FOX,
             A. ; CHEVALIER, P. ; CHANTRE, A. ; MEISTER, T. F. ; AUFINGER,
             K. ; SCHROTER, M.: Advanced process modules and architectures
             for half-terahertz SiGe:C HBTs. In: *IEEE Bipolar/BiCMOS Circuits
             and Technology Meeting (BCTM)*, 2009, pp. 9–16

[Fis12]   FISCHER, G.: Substrate Coupling for 500GHz HBTs. In: *Bipolar
             Arbeitskreis, Munich, Germany*, 2012

[Fox10]   FOX, A.: IHP Testchip T246 / IHP Microelectronics. 2010. –
             Forschungsbericht

[Fri02]   FRIEDRICH, M.: *Ein analytisches Modell zur Simulation von Silizium-
             Germanium Heterojunction-Bipolartransistoren in integrierten Schal-
             tungen*, Ruhr-Universität Bochum, Diss., 2002

[GDLMD67]   GHOSH, H.N. ; DE LA MONEDA, F.H. ; DONO, N.R.: Computer-aided
             transistor design, characterization, and optimization. In: *Proceedings
             of the IEEE* 55 (1967), no. 11, pp. 1897–1912. – ISSN 0018–9219

[GG76]   GOVER, A. ; GAASH, A.: Experimental model aid for planar design
             of transistor characteristics in integrated circuits . In: *Solid-State
             Electronics* 19 (1976), no. 2, pp. 125 – 127. – ISSN 0038–1101

[GR98]   GABL, R. ; REISCH, M.: Emitter series resistance from open-collector
             measurements-influence of the collector region and the parasitic pnp
             transistor. In: *IEEE Transactions on Electron Devices* 45 (1998), no.
             12, pp. 2457–2465. – ISSN 0018–9383

[GTB97]   GOBERT, Y. ; TASKER, P.J. ; BACHEM, K. H.: A physical, yet simple,
             small-signal equivalent circuit for the heterojunction bipolar transis-
             tor. In: *IEEE Transactions on Microwave Theory and Techniques* 45
             (1997), no. 1, pp. 149–153. – ISSN 0018–9480

[Gum64]   GUMMEL, H.K.: A self-consistent iterative scheme for one-dimensional
             steady state transistor calculations. In: *IEEE Transactions on Elec-
             tron Devices* 11 (1964), no. 10, pp. 455–465. – ISSN 0018–9383

[HBB⁺10]   HEINEMANN, B. ; BARTH, R. ; BOLZE, D. ; DREWS, J. ; FISCHER, G.G. ; FOX, A. ; FURSENKO, O. ; GRABOLLA, T. ; HAAK, U. ; KNOLL, D. ; KURPS, R. ; LISKER, M. ; MARSCHMEYER, S. ; RÜCKER, H. ; SCHMIDT, D. ; SCHMIDT, J. ; SCHUBERT, M.A. ; TILLACK, B. ; WIPF, C. ; WOLANSKY, D. ; YAMAMOTO, Y.: SiGe HBT technology with fT/fmax of 300GHz/500GHz and 2.0 ps CML gate delay. In: *IEEE International Electron Devices Meeting (IEDM)*, 2010, pp. 30.5.1–30.5.4. – ISSN 0163–1918

[HBK⁺06]   HEINEMANN, B. ; BARTH, R. ; KNOLL, D. ; RÜCKER, H. ; TILLACK, B. ; WINKLER, W.: High-Performance BiCMOS Technologies without Epitaxially-Buried Subcollectors and Deep Trenches. In: *Third International SiGe Technology and Device Meeting (ISTDM)*, 2006, pp. 1–2

[HC14]   HUSZKA, Z. ; CHAKRAVORTY, A.: Implementation of Delay-Time-Based Nonquasi-Static Bipolar Transistor Models in Circuit Simulators. In: *IEEE Transactions on Electron Devices* 61 (2014), no. 8, pp. 3004–3006. – ISSN 0018–9383

[HCS11a]   HUSZKA, Z. ; CELI, D. ; SEEBACHER, E.: A Novel Low-Bias Charge Concept for HBT/BJT Models Including Heterobandgap and Temperature Effects - Part I: Theory. In: *IEEE Transactions on Electron Devices* 58 (2011), no. 2, pp. 348 –356. – ISSN 0018–9383

[HCS11b]   HUSZKA, Z. ; CELI, D. ; SEEBACHER, E.: A Novel Low-Bias Charge Concept for HBT/BJT Models Including Heterobandgap and Temperature Effects - Part II: Implementation, Parameter Extraction and Verification. In: *IEEE Transactions on Electron Devices* 58 (2011), no. 2, pp. 357 –363. – ISSN 0018–9383

[HJ08]   HONG, S.-M. ; JUNGEMANN, C.: Deterministic simulation of SiGe HBTs based on the Boltzmann equation. In: *38th European Solid-State Device Research Conference (ESSDERC)*, 2008, pp. 170–173. – ISSN 1930–8876

[HJ09]   HONG, S.-M. ; JUNGEMANN, C.: Electron transport in extremely scaled SiGe HBTs. In: *IEEE Bipolar/BiCMOS Circuits and Technology Meeting (BCTM)*, 2009, pp. 67–74. – ISSN 1088–9299

[HJJ⁺07]   HADZIABDIC, D. ; JIANG, C. ; JOHANSEN, T.K. ; FISCHER, G.G. ; HEINEMANN, B. ; KROZER, V.: De-embedding and modelling of pnp SiGe HBTs. In: *European Microwave Integrated Circuit Conference (EuMIC)*, 2007, pp. 195–198

[HRB⁺02]   HEINEMANN, B. ; RÜCKER, H. ; BARTH, R. ; BAUER, J. ; BOLZE, D. ; BUGIEL, E. ; DREWS, J. ; EHWALD, K. E. ; GRABOLLA, T. ; HAAK, U. ;

HOPPNER, W. ; KNOLL, D. ; KRUGER, D. ; KUCK, B. ; KURPS, R. ;
MARSCHMEYER, M. ; RICHTER, H.H. ; SCHLEY, P. ; SCHMIDT, D. ;
SCHOLZ, R. ; TILLACK, B. ; WINKLER, W. ; WOLNSKY, D. ; WULF,
H.-E. ; YAMAMOTO, Y. ; ZAUMSEIL, P.: Novel collector design for
high-speed SiGe:C HBTs. In: *International Electron Devices Meeting
(IEDM)*, 2002, pp. 775–778

[HS09]   HUSZKA, Z. ; SEEBACHER, E.: Extraction of RE and its temperature dependence from RF measurements. In: *9th HICUM Workshop,
Würzburg, Germany*, 2009

[HSVA06]   HASAN, S. ; SALAHUDDIN, Sayeef ; VAIDYANATHAN, M. ; ALAM, M.A.:
High-frequency performance projections for ballistic carbon-nanotube
transistors. In: *IEEE Transactions on Nanotechnology* 5 (2006), no.
1, pp. 14–22. – ISSN 1536–125X

[Hus]   HUSZKA, Z. *private communication*

[JDSC10]   JACOB, J. ; DASGUPTA, A. ; SCHROTER, M. ; CHAKRAVORTY, A.:
Modeling Nonquasi-Static Effects in SiGe HBTs. In: *IEEE Transactions on Electron Devices* 57 (2010), no. 7, pp. 1559–1566. – ISSN
0018–9383

[JM05]   JUNGEMANN, C. ; MEINERZHAGEN, B.: Über die Vernachlässigung von
Beschleunigungseffekten in der Simulation von Siliziumbauelementen.
In: *ANALOG - 8. GMM/ITG-Diskussionssitzung: Entwicklung von
Analogschaltungen mit CAE-Methoden*. Hannover, Deutschland, 2005

[JNM01]   JUNGEMANN, C. ; NEINHUS, B. ; MEINERZHAGEN, B.: Comparative
study of electron transit times evaluated by DD, HD, and MC device simulation for a SiGe HBT. In: *IEEE Transactions on Electron
Devices* 48 (2001), no. 10, pp. 2216–2220. – ISSN 0018–9383

[JR11]   JAIN, V. ; RODWELL, M.J.W.: Transconductance Degradation in
Near-THz InP Double-Heterojunction Bipolar Transistors. In: *IEEE
Electron Device Letters* 32 (2011), no. 8, pp. 1068–1070. – ISSN
0741–3106

[Jun14]   JUNGEMANN, C.: Validity of Macroscopic Noise Models in the case
of High-Frequency Bipolar Transistors. In: *International Conference
on Simulation of Semiconductor Processes and Devices (SISPAD);
Workshop "Compact Modeling - Enabling Better Insight of Device Features"*. Yokohama, Japan, 2014

[KH75]   KUMAR, R. ; HUNTER, L.P.: Collector capacitance and high-level
injection effects in bipolar transistors. In: *IEEE Transactions on
Electron Devices* 22 (1975), no. 2, pp. 51–60. – ISSN 0018–9383

[Kir62]   KIRK, C.T.:   A theory of transistor cutoff frequency (fT) falloff at high current densities. In: *IRE Transactions on Electron Devices* 9 (1962), no. 2, pp. 164–174. – ISSN 0096–2430

[KLU+10]   KAZIOR, T.E. ; LAROCHE, J.R. ; URTEAGA, M. ; BERGMAN, J. ; CHOE, M.J. ; LEE, K.J. ; SEONG, T. ; SEO, M. ; YEN, A. ; LUBYSHEV, D. ; FASTENAU, J.M. ; LIU, W.K. ; SMITH, D. ; CLARK, D. ; THOMPSON, R. ; BULSARA, M.T. ; FITZGERALD, E.A. ; DRAZEK, C. ; GUIOT, E.: High Performance Mixed Signal Circuits Enabled by the Direct Monolithic Heterogeneous Integration of InP HBT and Si CMOS on a Silicon Substrate. In: *IEEE Compound Semiconductor Integrated Circuit Symposium (CSICS)*, 2010, pp. 1–4. – ISSN 1550–8781

[Kra15]   KRAUSE, J.: *Model Parameter Extraction for very Advanced Heterojunction Bipolar Transistors (to be published)*, Technische Universität Dresden, Diss., 2015

[KS14]   KRAUSE, J. ; SCHROTER, M.: An evaluation of methods for determining the emitter resistance of SiGe HBTs. In: *unpublished* (2014)

[KWSV90]   KUNDERT, K. ; WHITE, J. ; SANGIOVANNI-VINCENTELLI, A.: *Steady-State Methods for Simulating Analog and Microwave Circuits*. Boston : Kluwer Academic Publishers, 1990

[LCN+03]   LIANG, Q. ; CRESSLER, J.D. ; NIU, G. ; LU, Y. ; FREEMAN, G. ; AHLGREN, D.C. ; MALLADI, R.M. ; NEWTON, K. ; HARAME, D.L.: A simple four-port parasitic deembedding methodology for high-frequency scattering parameter and noise characterization of SiGe HBTs. In: *IEEE Transactions on Microwave Theory and Techniques* 51 (2003), no. 11, pp. 2165–2174. – ISSN 0018–9480

[LYS14]   LIU, J. ; YU, Z. ; SUN, L.: A Broadband Model Over 1-220 GHz for GSG Pad Structures in RF CMOS. In: *IEEE Electron Device Letters* 35 (2014), no. 7, pp. 696–698. – ISSN 0741–3106

[LZPS14]   LEHMANN, S. ; ZIMMERMANN, Y. ; PAWLAK, A. ; SCHROTER, M.: Characterization of the Static Thermal Coupling Between Emitter Fingers of Bipolar Transistors. In: *IEEE Transactions on Electron Devices* 61 (2014), no. 11, pp. 3676–3683. – ISSN 0018–9383

[Mas54]   MASON, S.J.: Power Gain in Feedback Amplifier. In: *Transactions of the IRE, Professional Group on Circuit Theory* CT-1 (1954), no. 2, pp. 20–25. – ISSN 0197–6389

[MP14]   M., Schröter ; PAWLAK, A.: Physics-based nonlinear compact modeling of HBTs for mm-wave applications. In: *International Microwave Symposium*. Tampa Bay, Florida, USA, 2014

[NKS14]   NARDMANN, T. ; KRAUSE, J. ; SCHROTER, M.: An evaluation of extraction methods for the emitter resistance for InP DHBTs. In: *IEEE Compound Semiconductor Integrated Circuit Symposium (CSICS)*, 2014, pp. 1–4

[NSC+13]   NARDMANN, T. ; SAKALAS, P. ; CHEN, F. ; ROSENBAUM, T. ; SCHROTER, M.: A Geometry Scalable Approach to InP HBT Compact Modeling for mm-Wave Applications. In: *IEEE Compound Semiconductor Integrated Circuit Symposium (CSICS)*, 2013, pp. 1–4

[NT84]   NING, T.H. ; TANG, D.D.: Method for determining the emitter and base series resistances of bipolar transistors. In: *IEEE Transactions on Electron Devices* 31 (1984), no. 4, pp. 409–412. – ISSN 0018–9383

[NV76]   NAKHLA, M. ; VLACH, J.: A piecewise harmonic balance technique for determination of periodic response of nonlinear systems. In: *IEEE Transactions on Circuits and Systems* 23 (1976), no. 2, pp. 85–91. – ISSN 0098–4094

[Paw08]   PAWLAK, A.: *A HICUM Model for 500 GHz SiGe Heterojunction Bipolar Transistors (in German)*, Technische Universität Dresden, Diplomarbeit, December 2008

[Paw11]   PAWLAK, A.: Modeling the electric field at the BC-junction of a BJT / TUD, CEDIC. 2011 (Version 1). – Forschungsbericht

[Paw13]   PAWLAK, A.: Test chip description for first wafer run in Dotseven / TUD, CEDIC. 2013. – Forschungsbericht

[Paw14]   PAWLAK, A.: Non-linear RF-behavior of HICUM/L2 / TUD, CEDIC. 2014 (Version 12). – Forschungsbericht

[PJ01]   PESIC, T. ; JANKOVIC, N.: An analytical model of the inverse base width modulation effect in SiGe graded heterojunction bipolar transistors . In: *Microelectronics Journal* 32 (2001), no. 9, pp. 713 – 718. – ISSN 0026–2692

[PKH01]   PAASSCHENS, J. ; KLOOSTERMAN, W.J. ; HAVENS, R.J.: Modelling two SiGe HBT specific features for circuit simulation. In: *Bipolar/BiCMOS Circuits and Technology Meeting (BCTM)*, 2001, pp. 38–41

[PLS14]   PAWLAK, A. ; LEHMANN, S. ; SCHROTER, M.: A Simple and Accurate Method for Extracting the Emitter and Thermal Resistance of BJTs and HBTs. In: *IEEE Bipolar/BiCMOS Circuits and Technology Meeting (BCTM)*, 2014, pp. 175–178. – ISSN 1088–9299

[PMPK02] PALESTRI, P. ; MASTRAPASQUA, M. ; PACELLI, A. ; KING, C.A.: A drift-diffusion/Monte Carlo simulation methodology for Si1-x Gex HBT design. In: *IEEE Transactions on Electron Devices* 49 (2002), no. 7, pp. 1242–1249. – ISSN 0018–9383

[PRH96] PFOST, M. ; REIN, H.-M. ; HOLZWARTH, T.: Modeling substrate effects in the design of high-speed Si-bipolar ICs. In: *IEEE Journal of Solid-State Circuits* 31 (1996), no. 10, pp. 1493–1501. – ISSN 0018–9200

[Pri58] PRITCHARD, R.L.: Two-Dimensional Current Flow in Junction Transistors at High Frequencies. In: *Proceedings of the IRE* 46 (1958), no. 6, pp. 1152–1160. – ISSN 0096–8390

[PS14a] PAWLAK, A. ; SCHROTER, M.: An Improved Transfer Current Model for RF and mm-Wave SiGe(C) Heterojunction Bipolar Transistors. In: *IEEE Transactions on Electron Devices* 61 (2014), no. 8, pp. 2612–2618. – ISSN 0018–9383

[PS14b] PAWLAK, A. ; SCHROTER, M.: Analytical model of weight factor hf0 / TUD, CEDIC. 2014 (Version 1). – Forschungsbericht

[PSF13] PAWLAK, A. ; SCHROTER, M. ; FOX, A.: Geometry scalable model parameter extraction for mm-wave SiGe-heterojunction transistors. In: *IEEE Bipolar/BiCMOS Circuits and Technology Meeting (BCTM)*, 2013, pp. 127–130. – ISSN 1088–9299

[PSK09a] PAWLAK, A. ; SCHRÖTER, M. ; KRAUSE, J.: A HICUM extension for medium current densities. In: *9th HICUM Workshop, Würzburg, Germany*, 2009

[PSK+09b] PAWLAK, A. ; SCHROTER, M. ; KRAUSE, J. ; WEDEL, G. ; SCHROTER, M. ; JUNGEMANN, C.: On the Feasibility of 500 GHz Silicon-Germanium HBTs. In: *International Conference on Simulation of Semiconductor Processes and Devices (SISPAD)*, 2009, pp. 1–4. – ISSN 1946–1569

[PSK+11] PAWLAK, A. ; SCHROTER, M. ; KRAUSE, J. ; CELI, D. ; DERRIER, N.: HICUM/2 v2.3 parameter extraction for advanced SiGe-heterojunction bipolar transistors. In: *IEEE Bipolar/BiCMOS Circuits and Technology Meeting (BCTM)*, 2011, pp. 195–198. – ISSN 1088–9299

[PSMK10] PAWLAK, A. ; SCHRÖTER, M. ; MUKHERJEE, A. ; KESSLER, T.: HICUM/L2 v2.30 overview. In: *10th HICUM Workshop, Dresden, Germany*, 2010

[RAH11]   RAYA, C. ; ARDOUIN, B. ; HUSZKA, Z.: Improving parasitic emitter resistance determination methods for advanced SiGe:C HBT transistors. In: *IEEE Bipolar/BiCMOS Circuits and Technology Meeting (BCTM)*, 2011, pp. 191–194. – ISSN 1088–9299

[Rei77]   REIN, H.-M.: Improving the large-signal models of bipolar transistors by dividing the intrinsic base into two lateral sections. In: *Electronics Letters* 13 (1977), no. 2, pp. 40–41. – ISSN 0013–5194

[Rei84]   REIN, H.-M.: A simple method for separation of the internal and external (peripheral) currents of bipolar transistors . In: *Solid-State Electronics* 27 (1984), no. 7, pp. 625 – 631. – ISSN 0038–1101

[RH12]   RÜCKER, H. ; HEINEMANN, B.: SiGe BiCMOS technology for mm-wave systems. In: *International SoC Design Conference (ISOCC)*, 2012, pp. 266–268

[RHB+03]   RÜCKER, H. ; HEINEMANN, B. ; BARTH, R. ; BOLZE, D. ; DREWS, J. ; HAAK, U. ; HOPPNER, W. ; KNOLL, D. ; KOPKE, K. ; MARSCHMEYER, S. ; RICHTER, H.H. ; SCHLEY, P. ; SCHMIDT, D. ; SCHOLZ, R. ; TILLACK, B. ; WINKLER, W. ; WULF, H.-E. ; YAMAMOTO, Y.: SiGe:C BiCMOS technology with 3.6 ps gate delay. In: *IEEE International Electron Devices Meeting (IEDM)*, 2003, pp. 5.3.1–5.3.4

[RHF12]   RÜCKER, H. ; HEINEMANN, B. ; FOX, A.: Half-Terahertz SiGe BiCMOS technology. In: *12th IEEE Topical Meeting on Silicon Monolithic Integrated Circuits in RF Systems (SiRF)*, 2012, pp. 133–136

[RKP+07]   RAYA, C. ; KAUFFMANN, N. ; POURCHON, F. ; CELI, D. ; ZIMMER, T.: Scalable approach for external collector resistance calculation. In: *IEEE International Conference on Microelectronic Test Structures (ICMTS)*, 2007, pp. 101–106

[RLB08]   RODWELL, M. ; LE, M. ; BRAR, B.: InP Bipolar ICs: Scaling Roadmaps, Frequency Limits, Manufacturable Technologies. In: *Proceedings of the IEEE* 96 (2008), no. 2, pp. 271–286. – ISSN 0018–9219

[RPZ+08]   RAYA, C. ; POURCHON, F. ; ZIMMER, T. ; CELI, D. ; CHEVALIER, P.: Scalable Approach for HBT's Base Resistance Calculation. In: *IEEE Transactions on Semiconductor Manufacturing* 21 (2008), no. 2, pp. 186–194. – ISSN 0894–6507

[RS91]   REIN, H.-M. ; SCHRÖTER, M.: Experimental determination of the internal base sheet resistance of bipolar transistors under forward-bias conditions. In: *Solid-State Electronics* 34 (1991), no. 3, pp. 301–308. – ISSN 0038–1101

[RSPL13]   ROSENBAUM, T. ; SCHROTER, M. ; PAWLAK, A. ; LEHMANN, S.: Automated transit time and transfer current extraction for single transistor geometries. In: *IEEE Bipolar/BiCMOS Circuits and Technology Meeting (BCTM)*, 2013, pp. 25–28. – ISSN 1088–9299

[RSS+08]   RADISIC, V. ; SAWDAI, D. ; SCOTT, D. ; DEAL, W.R. ; DANG, Linh ; LI, D. ; CAVUS, A. ; FUNG, A. ; SAMOSKA, L. ; TO, R. ; GAIER, T. ; LAI, R.: Demonstration of 184 and 255-GHz Amplifiers Using InP HBT Technology. In: *IEEE Microwave and Wireless Components Letters* 18 (2008), no. 4, pp. 281–283. – ISSN 1531–1309

[SC10]   SCHROTER, M. ; CHAKRAVORTY, A.: *Compact Hierarchical Bipolar Transistor Modeling With HiCUM*. World Scientific Publishing Co., Inc., 2010. – ISBN 978–981–4273–21–3

[Sch00]   SCHROTER, M.: Methods for extracting parameters of geometry scalable compact bipolar transistor models / TUD, CEDIC. 2000. – Forschungsbericht

[Sch05]   SCHROTER, M.: High-Frequency Circuit Design Oriented Compact Bipolar Transistor Modeling with HICUM. In: *IEICE Transactions on Electronics* Vol.E88-C (2005), no. No.6, pp. 1098–1113

[Sch13]   SCHROTER, M.: Geometry scaling / TUD, CEDIC. 2013. – Forschungsbericht

[Sch14]   SCHROTER, M.: Compact modeling of neutral base recpombination in mm-wave SiGe HBTs / TUD, CEDIC. 2014 (Version 12). – Forschungsbericht

[SCJH00]   SALMON, S.L. ; CRESSLER, J.D. ; JAEGER, R.C. ; HARAME, D.L.: The influence of Ge grading on the bias and temperature characteristics of SiGe HBTs for precision analog circuits. In: *IEEE Transactions on Electron Devices* 47 (2000), no. 2, pp. 292–298. – ISSN 0018–9383

[SCS+12]   SCHROTER, M. ; CLAUS, M. ; SAKALAS, P. ; WANG, D. ; HAFERLACH, M.: An overview on the state-of-the-art of Carbon-based radio-frequency electronics. In: *IEEE Bipolar/BiCMOS Circuits and Technology Meeting (BCTM)*, 2012, pp. 1–8. – ISSN 1088–9299

[SCS14]   SHON, Y. ; CHAKRAVORTY, A. ; SCHRÖTER, M.: Small-Signal Modeling of the Lateral NQS Effect in SiGe HBTs. In: *IEEE Bipolar/BiCMOS Circuits and Technology Meeting (BCTM)*, 2014, pp. 203–206. – ISSN 1088–9299

[SFR93]   SCHROTER, M. ; FRIEDRICH, M. ; REIN, H.-M.: A generalized integral charge-control relation and its application to compact models for silicon-based HBT's. In: *IEEE Transactions on Electron Devices* 40 (1993), no. 11, pp. 2036–2046. – ISSN 0018–9383

[SHD+12]   STEIN, F. ; HUSZKA, Z. ; DERRIER, N. ; MANEUX, C. ; CELI, D.: Extraction of the emitter related space charge weighting factor parameters of HICUM L2.30 using the Lambert W function. In: *IEEE Bipolar/BiCMOS Circuits and Technology Meeting (BCTM)*, 2012, pp. 1–4. – ISSN 1088–9299

[Sie02]   SIEGEL, P.H.: Terahertz technology. In: *IEEE Transactions on Microwave Theory and Techniques* 50 (2002), no. 3, pp. 910–928. – ISSN 0018–9480

[SKAA13]   SCHROTER, M. ; KRAUSE, J. ; AUFINGER, K. ; ARDOUIN, B.: Test structures for SiGe HBT compact modeling, parameter extraction, and circuit design (version 10) / TUD, CEDIC. 2013. – Forschungsbericht

[SKR+11]   SCHROTER, M. ; KRAUSE, J. ; RINALDI, N. ; WEDEL, G. ; HEINEMANN, B. ; CHEVALIER, P. ; CHANTRE, A.: Physical and Electrical Performance Limits of High-Speed Si GeC HBTs – Part II: Lateral Scaling. In: *IEEE Transactions on Electron Devices* 58 (2011), no. 11, pp. 3697–3706. – ISSN 0018–9383

[SL99]   SCHROTER, M. ; LEE, T.-Y.: Physics-based minority charge and transit time modeling for bipolar transistors. In: *IEEE Transactions on Electron Devices* 46 (1999), no. 2, pp. 288–300. – ISSN 0018–9383

[SL07]   SCHROTER, M. ; LEHMANN, S.: The rectangular bipolar transistor tetrode structure and its application. In: *IEEE International Conference on Microelectronic Test Structures (ICMTS)*, 2007, pp. 206–209

[SMNJ04]   SHERIDAN, D.C. ; MURTY, R.M. ; NEWTON, K.M. ; JOHNSON, J.B.: Generation and integration of scalable bipolar compact models [HBT example]. In: *Bipolar/BiCMOS Circuits and Technology Meeting (BCTM)*, 2004, pp. 132–139

[SPL00]   SCHROTER, M. ; PEHLKE, D.R. ; LEE, T.-Y.: Compact modeling of high-frequency distortion in silicon integrated bipolar transistors. In: *IEEE Transactions on Electron Devices* 47 (2000), no. 7, pp. 1529–1535. – ISSN 0018–9383

[SPM13]   SCHROTER, M. ; PAWLAK, A. ; MUKHERJEE, A.: *HICUM / L2 - A geometry scalable physics-based compact bipolar transistor model, Documentation of model version 2.32*, August 2013

[SR95]   SCHROTER, M. ; REIN, H.-M.: Investigation of very fast and high-current transients in digital bipolar IC's using both a new compact model and a device simulator. In: *IEEE Journal of Solid-State Circuits* 30 (1995), no. 5, pp. 551–562. – ISSN 0018–9200

[SRR+99]   SCHROTER, M. ; REIN, H.-M. ; RABE, W. ; REIMANN, R. ; WASSENER,
           H.-J. ; KOLDEHOFF, A.: Physics- and process-based bipolar transistor
           modeling for integrated circuit design. In: *IEEE Journal of Solid-State
           Circuits* 34 (1999), no. 8, pp. 1136–1149. – ISSN 0018–9200

[SRVC14]   SCHROTER, M. ; ROSENBAUM, T. ; VOINIGESCU, S.P. ; CHEVALIER,
           P.: A TCAD-based roadmap for high-speed SiGe HBTs. In: *14th
           IEEE Topical Meeting on Silicon Monolithic Integrated Circuits in Rf
           Systems (SiRF)*, 2014, pp. 80–82

[ST04]   SCHRÖTER, M. ; TRAN, H.: Modeling of charge and collector field
           in Si-based bipolar transistors. In: *WCM, International NanoTech
           Meeting*, 2004, pp. 102–107

[Ste13]   STEIN, F.: Selected Topics in Bipolar Modeling and Measurement.
           In: *Bipolar Arbeitskreis, Frankfurt (Oder), Germany*, 2013

[SW96]   SCHROTER, M. ; WALKEY, D.J.: Physical modeling of lateral scaling
           in bipolar transistors. In: *IEEE Journal of Solid-State Circuits* 31
           (1996), no. 10, pp. 1484–1492. – ISSN 0018–9200

[SWH+11]   SCHROTER, M. ; WEDEL, G. ; HEINEMANN, B. ; JUNGEMANN, C. ;
           KRAUSE, J. ; CHEVALIER, P. ; CHANTRE, A.: Physical and Electrical
           Performance Limits of High-Speed SiGeC HBTs – Part I: Vertical
           Scaling. In: *IEEE Transactions on Electron Devices* 58 (2011), no.
           11, pp. 3687–3696. – ISSN 0018–9383

[SZ06]   SCHROTER, M. ; ZIMMERMANN, Y.: *TRADICA - Physics- and
           Process-Based Model Generation for Integrated Devices*. TUD,
           CEDIC, 2006

[THJB05]   TIEMEIJER, L.F. ; HAVENS, R.J. ; JANSMAN, A.B.M. ; BOUTTE-
           MENT, Y.: Comparison of the "pad-open-short" and "open-short-load"
           deembedding techniques for accurate on-wafer RF characterization of
           high-quality passives. In: *IEEE Transactions on Microwave Theory
           and Techniques* 53 (2005), no. 2, pp. 723–729. – ISSN 0018–9480

[TPK12]   VAN DER TOORN, R. ; PAASSCHENS, J.C.J. ; KLOOSTERMAN, W.J.:
           The Mextram Bipolar Transistor Model, level 504.11.0, Dec. 6 2012

[TSW+97]   TRAN, H. ; SCHROTER, M. ; WALKEY, D.J. ; MARCHESAN, D. ;
           SMY, T.J.: Simultaneous extraction of thermal and emitter series
           resistances in bipolar transistors. In: *Bipolar/BiCMOS Circuits and
           Technology Meeting (BCTM)*, 1997, pp. 170–173. – ISSN 1088–9299

[VDA+12]   VOINIGESCU, S.P. ; DACQUAY, E. ; ADINOLFI, V. ; SARKAS, I. ; BAL-
           TEANU, A. ; TOMKINS, A. ; CELI, D. ; CHEVALIER, P.: Charac-
           terization and Modeling of an SiGe HBT Technology for Transceiver

Applications in the 100 - 300-GHz Range. In: *IEEE Transactions on Microwave Theory and Techniques* 60 (2012), no. 12, pp. 4024–4034. – ISSN 0018–9480

[Ver91]   VERSLEIJEN, M.P.J.G.: Distributed high frequency effects in bipolar transistors. In: *Bipolar Circuits and Technology Meeting (BCTM)*, 1991, pp. 85–88

[WCW87]   VAN WIJNEN, P.J. ; CLAESSEN, H.R. ; WOLSHEIMER, E.A.: A new straightforward calibration and correction procedure for "on wafer" high-frequency s-parameter measurements (45 MHz-18 GHz). In: *IEEE Bipolar/BiCMOS Circuits and Technology Meeting (BCTM)*, 1987

[Web54]   WEBSTER, W.M.: On the Variation of Junction-Transistor Current-Amplification Factor with Emitter Current. In: *Proceedings of the IRE* 42 (1954), no. 6, pp. 914–920. – ISSN 0096–8390

[Wed]   WEDEL, G. *private communication*

[WS10]   WEDEL, G. ; SCHROTER, M.: Hydrodynamic simulations for advanced SiGe HBTs. In: *IEEE Bipolar/BiCMOS Circuits and Technology Meeting (BCTM)*, 2010, pp. 237–244. – ISSN 1088–9299

[WWS+14]   WILLIAMS, T. ; WEE, B. ; SAINI, R. ; MATHIAS, S. ; BOSSCHE, M.V.: A digital, PXI-based active load-pull tuner to maximize throughput of a load-pull test bench. In: *83rd ARFTG Microwave Measurement Conference*, 2014, pp. 1–4

[Zim15]   ZIMMERMANN, Y.: *Yet untitled*, Technische Universität Dresden, Diss., 2015

A | Modeling

A.1 Derivatives of the diffusion charge

The diffusion charge in HICUM/L2 consists of three components, each of which is discussed separately in this section.

The emitter charge is given by

$$\Delta Q_{\text{Ef}} = \frac{\tau_{\text{Ef}} I_{\text{Tf}}}{1 + g_{\tau\text{fE}}}, \tag{A.1}$$

with

$$\tau_{\text{Ef}} = \tau_{\text{Ef0}} \left(\frac{I_{\text{Tf}}}{I_{\text{CK}}} \right)^{g_{\tau\text{fE}}}. \tag{A.2}$$

The derivative of Q_{Ef} with respect to I_{Tf} is

$$\frac{\mathrm{d}\Delta Q_{\text{Ef}}}{\mathrm{d}I_{\text{Tf}}} = \tau_{\text{Ef}}, \tag{A.3}$$

while the derivative with respect to I_{CK} is

$$\frac{\mathrm{d}\Delta Q_{\text{Ef}}}{\mathrm{d}I_{\text{CK}}} = -\frac{g_{\tau\text{fE}}}{1 + g_{\tau\text{fE}}} \tau_{\text{Ef}} \left(\frac{I_{\text{Tf}}}{I_{\text{CK}}} \right). \tag{A.4}$$

From

$$-\frac{\mathrm{d}\Delta Q_{\text{Ef}}}{\mathrm{d}I_{\text{CK}}} > \frac{\mathrm{d}\Delta Q_{\text{Ef}}}{\mathrm{d}I_{\text{Tf}}}, \tag{A.5}$$

which is the condition for the negative (correct) $\Im\{Y_{12}\}$ follows

$$\frac{g_{\tau\text{fE}}}{1 + g_{\tau\text{fE}}} \tau_{\text{Ef}} \left(\frac{I_{\text{Tf}}}{I_{\text{CK}}} \right) > \tau_{\text{Ef}} \tag{A.6}$$

and

$$\frac{I_{\text{Tf}}}{I_{\text{CK}}} > 1 + \frac{1}{g_{\tau\text{fE}}}. \tag{A.7}$$

Therefore, for all currents large enough to fulfill the condition above the resulting $\mathrm{d}\Delta Q_{\text{Ef}}/\mathrm{d}V_{\text{C'E'}}$ and therefore the corresponding influence on $\Im\{Y_{12}\}$ is negative.

The actual implementation of the barrier voltage is in the form $\Delta V_{\mathrm{cBar}} = f(I_{\mathrm{Tf}} - I_{\mathrm{CK}})$. For the derivatives follow

$$\frac{\mathrm{d}\Delta V_{\mathrm{cBar}}}{\mathrm{d}I_{\mathrm{Tf}}} = -\frac{\mathrm{d}\Delta V_{\mathrm{cBar}}}{\mathrm{d}I_{\mathrm{CK}}}. \tag{A.8}$$

The base charge corresponding to the barrier is calculated as

$$\Delta Q_{\mathrm{Bf,b}} = \tau_{\mathrm{Bvl}}I_{\mathrm{Tf}}\left[\exp\left(\frac{\Delta V_{\mathrm{cBar}}}{V_{\mathrm{T}}}\right) - 1\right]. \tag{A.9}$$

The corresponding derivatives are

$$\frac{\mathrm{d}\Delta Q_{\mathrm{Bf,b}}}{\mathrm{d}I_{\mathrm{Tf}}} = \frac{\Delta Q_{\mathrm{Bf,b}}}{I_{\mathrm{Tf}}} + \frac{\tau_{\mathrm{Bvl}}I_{\mathrm{Tf}}}{V_{\mathrm{T}}}\exp\left(\frac{\Delta V_{\mathrm{cBar}}}{V_{\mathrm{T}}}\right)\frac{\mathrm{d}\left(\Delta V_{cBar}\right)}{\mathrm{d}I_{\mathrm{Tf}}}, \tag{A.10}$$

and

$$\frac{\mathrm{d}\Delta Q_{\mathrm{Bf,b}}}{\mathrm{d}I_{\mathrm{CK}}} = -\frac{\tau_{\mathrm{Bvl}}I_{\mathrm{Tf}}}{V_{\mathrm{T}}}\exp\left(\frac{\Delta V_{\mathrm{cBar}}}{V_{\mathrm{T}}}\right)\frac{\mathrm{d}\Delta V_{\mathrm{cBar}}}{\mathrm{d}I_{\mathrm{Tf}}}. \tag{A.11}$$

From both derivatives follows

$$\frac{\mathrm{d}\Delta Q_{\mathrm{Bf,b}}}{\mathrm{d}I_{\mathrm{Tf}}} + \frac{\mathrm{d}\Delta Q_{\mathrm{Bf,b}}}{\mathrm{d}I_{\mathrm{CK}}} = \frac{\Delta Q_{\mathrm{Bf,b}}}{I_{\mathrm{Tf}}}, \tag{A.12}$$

which is always positive. Therefore, caused by the actual implementation of this component the resulting impact on $\Im\{Y_{12}\}$ will always be positive.

The Kirk-effect related charge component reads

$$\Delta Q_{\mathrm{fh}} = \tau_{\mathrm{hCs}}I_{\mathrm{Tf}}w^2\exp\left(\frac{\Delta V_{\mathrm{cBar}} - V_{\mathrm{cBar}}}{V_{\mathrm{T}}}\right), \tag{A.13}$$

with the derivatives

$$\frac{\mathrm{d}\Delta Q_{\mathrm{fh}}}{\mathrm{d}I_{Tf}} = \Delta Q_{\mathrm{fh}}\left[\frac{1}{I_{\mathrm{Tf}}} + \frac{2}{w}\frac{\mathrm{d}w}{\mathrm{d}I_{\mathrm{Tf}}} + \frac{1}{V_{\mathrm{T}}}\frac{\mathrm{d}\Delta V_{\mathrm{cBar}}}{\mathrm{d}I_{\mathrm{Tf}}}\right], \tag{A.14}$$

and

$$\frac{\mathrm{d}\Delta Q_{\mathrm{fh}}}{\mathrm{d}I_{\mathrm{CK}}} = \Delta Q_{\mathrm{fh}}\left[\frac{2}{w}\frac{\mathrm{d}w}{\mathrm{d}I_{\mathrm{CK}}} - \frac{1}{V_{\mathrm{T}}}\frac{\mathrm{d}\Delta V_{\mathrm{cBar}}}{\mathrm{d}I_{\mathrm{Tf}}}\right]. \tag{A.15}$$

The derivatives of w are

$$\frac{\mathrm{d}w}{\mathrm{d}I_{\mathrm{Tf}}} = \frac{I_{\mathrm{CK}}w}{I_{\mathrm{Tf}}^2\sqrt{i^2 + a_{\mathrm{hC}}}} \tag{A.16}$$

and

$$\frac{\mathrm{d}w}{\mathrm{d}I_{\mathrm{CK}}} = -\frac{w}{I_{\mathrm{Tf}}\sqrt{i^2 + a_{\mathrm{hC}}}}. \tag{A.17}$$

Calculating the sum of both derivatives leads to

$$\frac{\mathrm{d}\Delta Q_{\mathrm{fh}}}{\mathrm{d}I_{\mathrm{Tf}}} + \frac{\mathrm{d}\Delta Q_{\mathrm{fh}}}{\mathrm{d}I_{\mathrm{CK}}} = \frac{\Delta Q_{\mathrm{fh}}}{I_{\mathrm{Tf}}}\left[1 + 2\frac{I_{\mathrm{CK}} - I_{\mathrm{Tf}}}{I_{\mathrm{Tf}}\sqrt{i^2 + a_{\mathrm{hC}}}}\right],\tag{A.18}$$

which has only a small positive overshoot and is becoming negative for sufficient large I_{Tf}.

A.2 Quasi-static modeling with the GICCR

A.2.1 Modeling voltage drops with the GICCR

In general, the transfer current of the GICCR is independent on the integration boundaries. The equivalence of integrating from the beginning to end of the complete transistor and moving the lower boundary to the beginning of the neutral base was shown in [Fri02], although with some unnecessary assumptions. The derivation there was restricted to the resistance of the neutral emitter. In the following it is shown that all DC voltage drops are always correctly covered by the GICCR. Furthermore, the correlation between the voltage drop (and the corresponding resistance) and the weight factors of the components from the weighted charge will be derived.

The GICCR master equation with variable integration boundaries reads

$$i_{\mathrm{T}} = \frac{qA_{\mathrm{E}}V_{\mathrm{T}}\overline{\mu_{\mathrm{nr}}n_{\mathrm{ir}}^2}}{\int_{x_1}^{x_u} h(x)p(x)\mathrm{d}x}\exp\left(\frac{V_{\mathrm{B'E'}}}{V_{\mathrm{T}}}\right)\left[\exp\left(-\frac{\varphi_{\mathrm{n}}(x_1)}{V_{\mathrm{T}}}\right) - \exp\left(-\frac{\varphi_{\mathrm{n}}(x_u)}{V_{\mathrm{T}}}\right)\right],\tag{A.19}$$

with the weight function (2.69) where in the following the assumption $\varphi_{\mathrm{n}}(x_u) \gg V_{\mathrm{T}}$ is made[1].

The complete weighted charge in the transistor reads

$$Q_{\mathrm{ph}} = qA_{\mathrm{E}}\int_{x_1}^{x_u} h(x)p(x)\mathrm{d}x.\tag{A.20}$$

The assumption that voltage drops are modeled correctly by the GICCR corresponds to

$$Q_{\mathrm{ph}}\exp\left(-\frac{\Delta\varphi_{\mathrm{n}}(x_{\mathrm{w}})}{V_{\mathrm{T}}}\right) = Q_{\mathrm{ph}} - q\int_{x_1}^{x_{\mathrm{w}}} h(x)p(x)\mathrm{d}x,\tag{A.21}$$

where x_{w} is an arbitrary location in the transistor and $\Delta\varphi_{\mathrm{n}}(x_{\mathrm{w}}) = \varphi_{\mathrm{n}}(x_{\mathrm{w}}) - \varphi_{\mathrm{n}}(x_1)$ is the voltage drop in the region between x_{w} and x_1. Note that by using φ_{n} rather that ψ this derivation is not limited to neutral regions.

From the DD transport equation, this voltage drop can be expressed by

$$\Delta\varphi_{\mathrm{n}}(x_{\mathrm{w}}) = \int_{x_1}^{x_{\mathrm{w}}} \frac{-j_{\mathrm{n}}(x)}{q\mu_{\mathrm{n}}(x)n(x)}\mathrm{d}x.\tag{A.22}$$

---

[1]These assumptions are not limiting the derived results.

Inserting

$$n(x) = \frac{n_{\mathrm{i}}(x)^2}{p(x)} \exp\left(\frac{\varphi_{\mathrm{p}}(x) - \varphi_{\mathrm{n}}(x)}{V_{\mathrm{T}}}\right) \tag{A.23}$$

and (2.69) into (A.22) returns

$$\Delta\varphi_{\mathrm{n}}(x_{\mathrm{w}}) = \int_{x_{\mathrm{l}}}^{x_{\mathrm{w}}} \frac{i_{\mathrm{T}} h(x) p(x) \exp\left(\frac{\varphi_{\mathrm{n}}(x)}{V_{\mathrm{T}}}\right)}{A_{\mathrm{E}} q \overline{\mu_{\mathrm{nr}} n_{\mathrm{ir}}^2} \exp\left(\frac{V_{\mathrm{B'E'}}}{V_{\mathrm{T}}}\right)} \mathrm{d}x. \tag{A.24}$$

Finally (A.19) is used to provide

$$\Delta\varphi_{\mathrm{n}}(x_{\mathrm{w}}) = V_{\mathrm{T}} \int_{x_{\mathrm{l}}}^{x_{\mathrm{w}}} \frac{q h(x) p(x)}{Q_{\mathrm{ph}} \exp\left(-\frac{\Delta\varphi_{\mathrm{n}}(x)}{V_{\mathrm{T}}}\right)} \mathrm{d}x, \tag{A.25}$$

with $\Delta\varphi_{\mathrm{n}}(x) = \varphi_{\mathrm{n}}(x) - \varphi_{\mathrm{n}}(x_{\mathrm{l}})$, that is inserted into (A.21) which then reads

$$\exp\left(-\int_{x_{\mathrm{l}}}^{x_{\mathrm{w}}} \frac{q h(x) p(x)}{Q_{\mathrm{ph}} \exp\left(-\frac{\Delta\varphi_{\mathrm{n}}(x)}{V_{\mathrm{T}}}\right)} \mathrm{d}x\right) = \frac{Q_{\mathrm{ph}} - \int_{x_{\mathrm{l}}}^{x_{\mathrm{w}}} h(x) p(x) \mathrm{d}x}{Q_{\mathrm{ph}}}. \tag{A.26}$$

Realizing that $\exp\left(\int_{x_{\mathrm{l}}}^{x_{\mathrm{u}}} f'(x)/f(x)\mathrm{d}x\right) = f(x_{\mathrm{u}})/f(x_{\mathrm{l}})$, above equation is only fulfilled when (A.21) is true verifying this equation.

From (A.21), the correspondence between the voltage drop in a specific region of the transistor and the weighted charge in this region can be derived. It is

$$\Delta Q_{\mathrm{ph}} = q A_{\mathrm{E}} \int_{x_{\mathrm{l}}}^{x_{\mathrm{w}}} h(x) p(x)\mathrm{d}x = Q_{\mathrm{ph}}\left[\exp\left(-\frac{\Delta\varphi_{\mathrm{n}}(x_{\mathrm{w}})}{V_{\mathrm{T}}}\right) - 1\right] \tag{A.27}$$

and

$$\Delta\varphi_{\mathrm{n}}(x_{\mathrm{w}}) = -V_{\mathrm{T}} \log\left(1 - \frac{\Delta Q_{\mathrm{ph}}}{Q_{\mathrm{ph}}}\right). \tag{A.28}$$

Since ΔQ_{ph} is a part of Q_{ph}, the latter equation can be interpreted as following. The voltage drop across regions in a transistor is large when the corresponding weighted charge is a large portion of the complete weighted charge.

A.3 Model implementation of the electric field at the base-collector junction

In this section, an overview of the model parameters, the employed model equations and the final implementation in Verilog-A for the electric field model at the BC junction are given.

A.3.1 Model parameters

A summary of all parameters for the model of E_{jC} is given in Tab. A.1.

Parameter	Description	Comment
V_{DCi}	built-in voltage of the base-collector SCR	already existing, extracted from $C_{jCi}(V_{BCi})$
V_{lim}	characteristic voltage of the field dependence of the electron velocity	already existing, extracted from I_{CK}
R_{Ci0}	low field resistance of the internal collector	already existing, extracted from I_{CK}
V_{PT}	punch-through voltage	already existing, extracted from I_{CK}
w_{Ci}	depth of the internal collector	
A_{E0}	area of the emitter window	
ε_r	relative permittivity	
$\tau_{CCe,00}$	low current transit time at $V_{B'C'} = 0\,\mathrm{V}$	
$\Delta\tau_{CCe,0}$	increase of the low current transit time	
$\Delta\tau_{CCe,PT}$	transit time increase in the ohmic region	
ζ_{Ohm}	fitting parameter for the transit time in the ohmic region	
ζ_{PT}	fitting parameter for the transit time in the punch-through region	
g_{jC}	fitting parameter for the transition to E_{inf}	

Table A.1: Employed parameters in the model for the electric field. Note, β_n which is extracted from I_{CK} is not included in the final model equations.

A.3.2 Model equations

The collector charge is given by

$$Q_{BC} = \varepsilon A_{E0} E_{jC}, \tag{A.29}$$

with the permittivity $\varepsilon = \varepsilon_r \varepsilon_0$ and the electric field model calculated as

$$E_{jC} = \begin{cases} E_{jC,Ohm}, & \text{for } I_T < I_{PT} \\ E_{jC,PT}, & \text{for } I_T \geq I_{PT} \end{cases}, \tag{A.30}$$

with the model equation of the ohmic case, reading

$$E_{jC,Ohm} = E_{jC0} - \frac{\tau_{CCe,0} I_{PT}}{A_E \varepsilon} \left[i_{Ohm} - \frac{i_{Ohm}^{\zeta_{Ohm}}}{\zeta_{Ohm}} \right] - \frac{\tau_{CCe,PT} I_{PT}}{A_E \varepsilon} \frac{i_{Ohm}^{\zeta_{Ohm}}}{\zeta_{Ohm}}, \tag{A.31}$$

with

$$i_{Ohm} = \frac{I_T}{I_{PT}}, \tag{A.32}$$

and the equation for the punch-through case

$$E_{jC,PT} = E_\infty + \frac{e_j + \sqrt{e_j + g_{jC} E_{lim}^2}}{2}. \tag{A.33}$$

It is

$$E_\infty = \frac{V_T}{w_{Ci}}, \tag{A.34}$$

and

$$e_j = -a_p i_{PT}^{\zeta_{PT}} + b_p i_{PT} + c_p, \tag{A.35}$$

with the coefficients

$$a_p = \frac{1}{2} \left(x_a + \sqrt{x_a^2} \right), \tag{A.36}$$

with

$$x_a = -\left(E_{CK} - E_{jC,PT} - E_\infty - \frac{1}{4} \frac{g_{jC} E_{lim}^2}{E_{jC,PT} - E_\infty} - b \right), \tag{A.37}$$

$$b_p = -2 \frac{\tau_{CCe,PT} (I_{CK} - I_{PT})}{\varepsilon A_{E0} \left(1 + c_p / \sqrt{c_p^2 + g_{jC} E_{lim}^2} \right)}, \tag{A.38}$$

and

$$c_p = E_{jC,PT} - E_\infty - \frac{1}{4} \frac{g_{jC} E_{lim}^2}{E_{jC,PT} - E_\infty}. \tag{A.39}$$

Note that the coefficients taken a limiting of a_p to 0 as well as the effects from the smoothing at high currents into account. The punch-through current is calculated as

$$i_{PT} = \frac{I_T - I_{PT}}{I_{CK} - I_{PT}}, \tag{A.40}$$

while the electric field correlated to high current effects is given by

$$E_{CK} = E_{lim} \frac{V_{Ci}/V_{lim}}{1 + V_{Ci}/V_{lim}}. \tag{A.41}$$

V_{Ci} is calculated from

$$V_{Ci} = V_{DCi} - V_{BCi}, \tag{A.42}$$

and

$$E_{lim} = \frac{V_{lim}}{w_{Ci}}. \tag{A.43}$$

The low current electric field is calculated as

$$E_{jC0} = E_{jC00} - \frac{Q_{jCi,0}}{\varepsilon A_{E0}} \tag{A.44}$$

with the low current electric field at $V_{B'C'} = 0\,\mathrm{V}$

$$E_{jC00} = \frac{Q_{jCi,bi}}{\varepsilon A_{E0}}, \tag{A.45}$$

the low current depletion charge from HICUM/L2 $Q_{jCi,0}$ and the depletion charge $Q_{jCi,bi}$ at $V_{B'C'} = V_{DCi}$.

The low current transit time is given by

$$\tau_{CCe,0} = \tau_{CCe,00} + \Delta\tau_{CCe,0}\,(c-1) + \Delta\tau_{CCe,0}\frac{V_{lim}}{V_{DCi}}\left(\frac{V_{DCi}}{V_{Ci,lim}}c - 1\right), \tag{A.46}$$

with

$$c = \frac{C_{jCi0}}{C_{jCi}}, \tag{A.47}$$

and the limited collector voltage

$$V_{Ci,lim} = V_{PT} - \frac{1}{2}\left[(V_{PT} - V_{Ci}) + \sqrt{(V_{PT} - V_{Ci})^2 + 0.01}\right]. \tag{A.48}$$

The current dividing both regions is given by

$$I_{PT,eq} = I_{lim}\left(k - \sqrt{k^2 - \frac{I_{CKl,\beta_n=1}}{I_{lim}} + \frac{V_{Ci}}{V_{PT}} + 0.01}\right), \tag{A.49}$$

with

$$k = \frac{I_{lim} + I_{CKl,\beta_n=1}}{2I_{lim}} - \frac{V_{Ci} + V_{lim}}{2V_{PT}}. \tag{A.50}$$

The term 0.01 is inserted for a more smooth curve at the peak of $I_{PT,eq}$. The limit for high V_{Ci} is implemented with the simple equation

$$I_{PT} = I_{PT,eq} + \sqrt{I_{PT,eq} + 0.01 I_{lim}^2}. \tag{A.51}$$

Using this equation a positive non-zero value of I_{PT} is ensured avoiding numerical issues for i_{Ohm}.

In above equations I_{lim} is given by

$$I_{lim} = \frac{V_{lim}}{R_{Ci0}}, \tag{A.52}$$

and $I_{CKl,\beta_n=1}$ by

$$I_{CKl,\beta_n=1} = I_{lim}\frac{E_{CK}}{E_{lim}}. \tag{A.53}$$

The punch-through transit time is given by

$$\tau_{\text{CCe,PT}} = \tau_{\text{CCe,0}} + \Delta\tau_{\text{PT}}\frac{I_{\text{PT}}}{I_{\text{lim}}}, \tag{A.54}$$

and the corresponding electric field by

$$E_{\text{jC,PT}} = E_{jC0} - \frac{\tau_{CCe,PT}I_{\text{PT}}}{A_E\varepsilon\zeta_{\text{Ohm}}}. \tag{A.55}$$

A.3.3 VA-code

In this section, the implementation in Verilog-A is given in the following listing.

```
parameter real wCi = 0;
parameter real AE = 1e-12;
parameter real epsr = 11.7;
parameter real tcce00 = 0;
parameter real dtcce0 = 0;
parameter real dtpt = 0;
parameter real zetaohm = 2;
parameter real zetapt = 2;
parameter real gjC = 1;

real eps, dummy;
real VCi, v_norm, vn_norm, VCi_lim;
real Ilim, IPT, IPT_eq, kPT, ICK0, tPT;
real iohm, iohm_zeta, ipt, ipt_zeta;
real tcce0;
real EjC00, EjC0, EjC, QjCi0, EjC_PT, ECK, Einf, Elim;
real a_poly, b_poly, c_poly;
real tCCe_it, CjCi_it;

if (wCi > 0) begin: Field_Mod
    eps = epsr*'P_EPS0;

    VCi = vdci_t-V(br_bici);

    'HICJQ(cjci0_t, vdci_t, zci, vptci_t, vdci_t, dummy, QjCi0
        )
    EjC00 = QjCi0/(eps*AE);
    EjC0 = EjC00-Qjci/(eps*AE);

    v_norm = VCi/vlim_t;
    vn_norm = v_norm/(1+v_norm);

    Ilim = vlim_t/rci0_t;
```

```
Elim = vlim_t/wCi;
ICK0 = Ilim*vn_norm;
ECK = (vlim_t/wCi)*vn_norm;

kPT = (Ilim+ICK0)/(2*Ilim)-(VCi+vlim_t)/(2*vpt);
IPT_eq = Ilim*(kPT-sqrt(kPT*kPT-ICK0/Ilim+VCi/vpt
    +0.01));
IPT = IPT_eq+sqrt(IPT_eq*IPT_eq+0.01*Ilim*Ilim);

cc = cjci0_t/Cjci;
VCi_lim = vpt-VCi;
VCi_lim = vpt-0.5*(VCi_lim+sqrt(VCi_lim*VCi_lim
    +0.01));
tcce0 = tcce00+dtcce0*(cc-1)+dtcce0*vlim_t/vdci_t*(
    vdci_t/VCi_lim*cc-1);
tPT = tcce0+dtpt*IPT/Ilim;

iohm = itf/IPT;
iohm_zeta = exp(zetaohm*ln(iohm));
ipt = (itf-IPT)/(ick-IPT);
ipt_zeta = exp(zetapt*ln(ipt));

EjC_PT = EjC0-tcce0*IPT/(AE*eps)*(1-1/zetaohm)-tPT*
    IPT/(AE*eps*zetaohm);
Einf = VT/wCi;

if (itf < IPT) begin
    c_poly = EjC0;
    b_poly = -tcce0*IPT/(AE*eps);
    a_poly = (tcce0-tPT)*IPT/(AE*eps*zetaohm);
    EjC = a_poly*iohm_zeta+b_poly*iohm+c_poly;
end else begin
    c_poly = EjC_PT-Einf-0.25*(gjC*Elim*Elim)/(
        EjC_PT-Einf);
    b_poly = -2*tPT*(ick-IPT)/(AE*eps*(1+c_poly/sqrt
        (c_poly*c_poly+gjC*Elim*Elim)));
    a_poly = -(Elim-EjC_PT-Einf-0.25*(gjC*Elim*Elim)
        /(EjC_PT-Einf)-b_poly);
    a_poly = 0.5*(a_poly+sqrt(a_poly*a_poly));
    EjC = -a_poly*ipt_zeta+b_poly*ipt+c_poly;
    EjC = Einf+(EjC+sqrt(EjC*EjC+gjC*Elim*Elim))/2;
end

Qjci = eps*AE*EjC;
```

end

Listing A.1: Verilog-A implementation of the field model.

B | Model verification

B.1 Verification of calibration methods

The comparison of the measured S-parameters from open and short deembedding structures using different calibration techniques (cf. 4.4.2.1) is shown in Fig. B.1. Only very small deviations are visible in the complete frequency range. The same holds for the transmission S-parameters, i.e. $\underline{S}_{12}$ and $\underline{S}_{21}$ (not shown in the pictures).

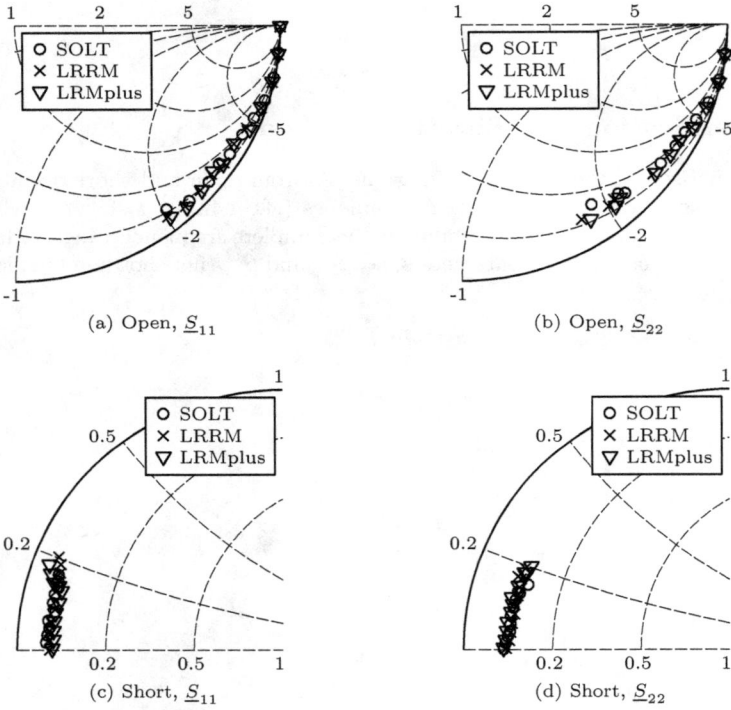

(a) Open, $\underline{S}_{11}$

(b) Open, $\underline{S}_{22}$

(c) Short, $\underline{S}_{11}$

(d) Short, $\underline{S}_{22}$

Figure B.1: Comparison of the different calibration methods in the frequency range [1 ... 110] GHz using passive deembedding structures.

B.2 Additional verification plots

In this section additional verification results of important characteristics not provided in chapter 4.4 are given.

B.2.1 DC results

In figures B.2 and B.3 output curves at fixed V_{BE} and fixed I_B are given. The maximum drive correlate to operating points slightly beyond peak f_T.

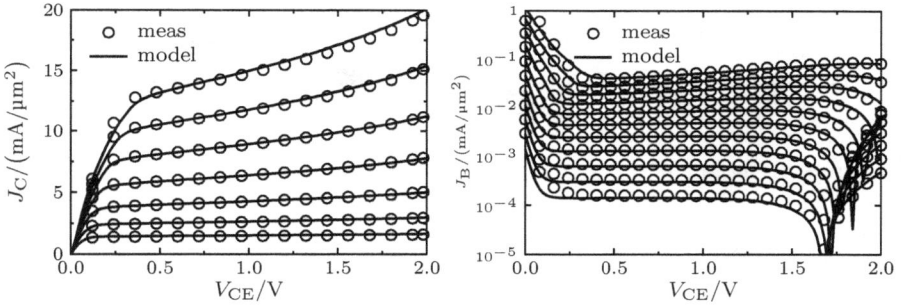

Figure B.2: Collector and base current from output curves at fixed base-emitter voltages $V_{BE} = [0.82 \ldots 0.94]\,V$.

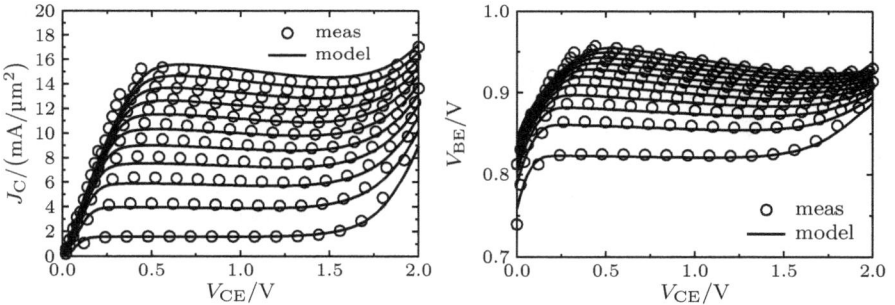

Figure B.3: Collector current and base-emitter voltage from output curves at fixed $I_B = [3 \ldots 60]\,\mu A$.

B.2.2 AC results

B.2.2.1 Derived quantities

In Fig. B.4, the terminal capacitances according to (4.3) are provided.

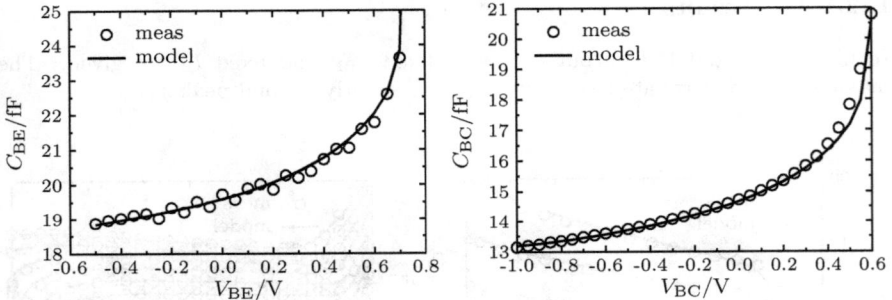

Figure B.4: Terminal BE- and BC-capacitance from (4.3) at spot-frequency $f = 5\,\mathrm{GHz}$.

B.2.2.2 Small-signal parameters

For small-signal parameters, different plot options are available providing more or less clear results. Utilizing transformed plots as the Smith Chart or polar plots reduce the number of plots by joining real and imaginary part or magnitude and phase into a single plot. However, they are less capable of highlighting actual inaccuracies. In this work real and imaginary part instead of magnitude and phase are plotted. Both are useful in certain operating ranges (roughly the former for low frequencies, the latter for high). However, the number of plots should be reduced. Thus, only one option is given.

The following plots for each Y-parameter are given as function of bias and frequency. For the former, $f = [1, 10, 20, 50, 50]\,\mathrm{GHz}$[1] for the 67 GHz setup and $f = [20, 100]\,\mathrm{GHz}$ for the 110 GHz setup is used as parameter. For the plots versus frequency, $J_C = [0.1, 1, 5, 10, 20]\,\mathrm{mA}/\mathrm{\mu m}^2$ for the 67 GHz setup and $J_C = [1, 10]\,\mathrm{mA}/\mathrm{\mu m}^2$ for the 110 GHz setup were used. Units in the legends of the following plots were omitted to reduce the size. The results are given in the figures B.5-B.8.

---

[1] Results at 67 GHz are more noisy compared to results at 50 GHz and do not provide any more insight. Thus, 50 GHz was used here.

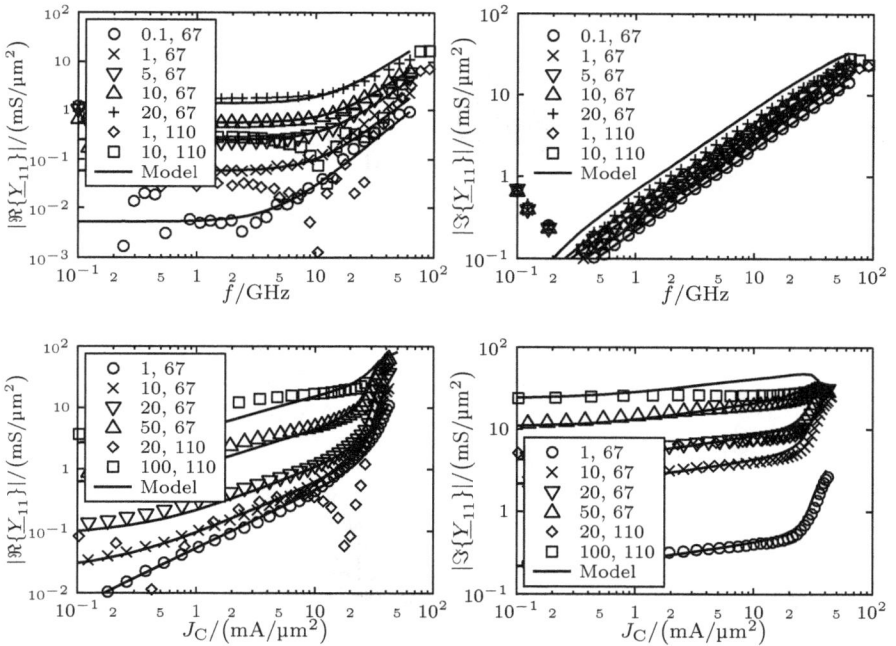

Figure B.5: Real and imaginary part of $\underline{Y}_{11}$.

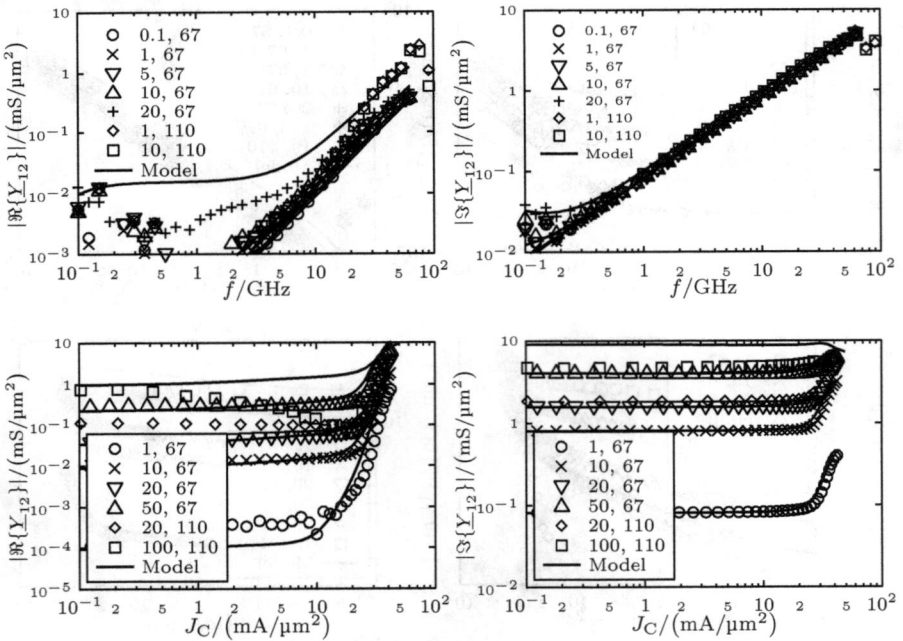

Figure B.6: Real and imaginary part of $\underline{Y}_{12}$.

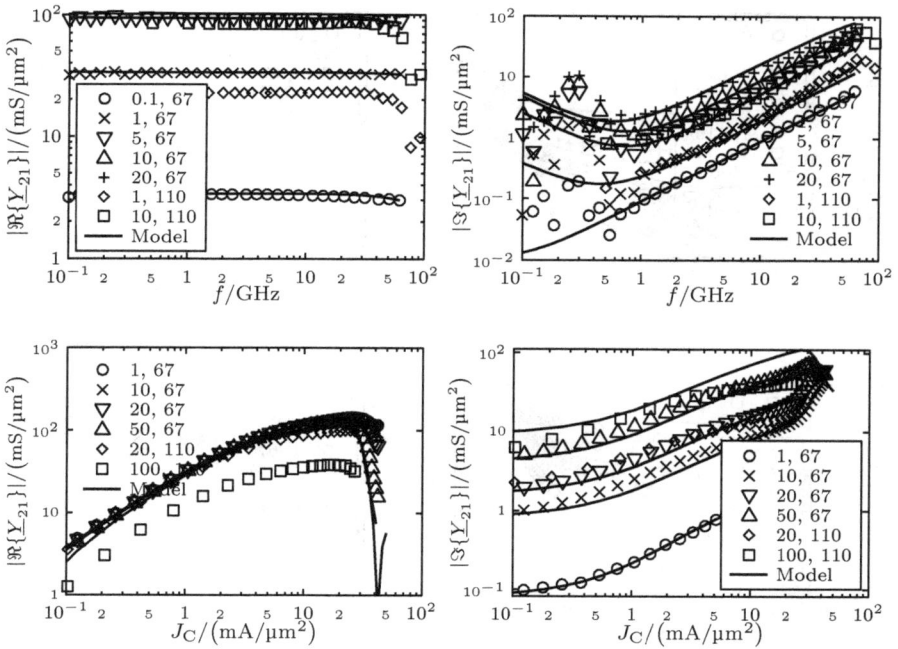

Figure B.7: Real and imaginary part of $\underline{Y}_{21}$.

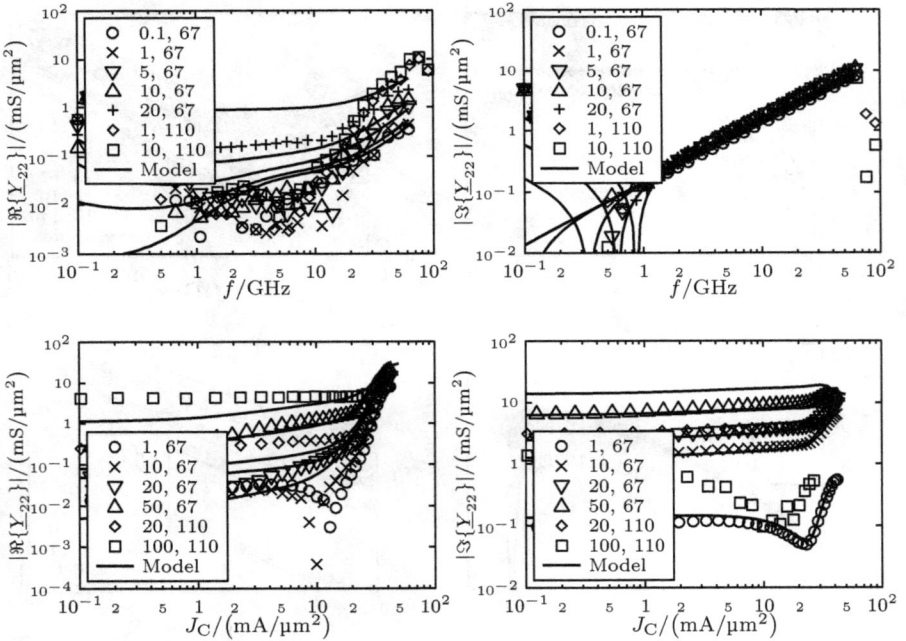

Figure B.8: Real and imaginary part of $\underline{Y}_{22}$.

B.2.3 Device scaling

Additionally to the plots in 4.4.4 figures B.9-B.12 provide the remaining important FOMs .

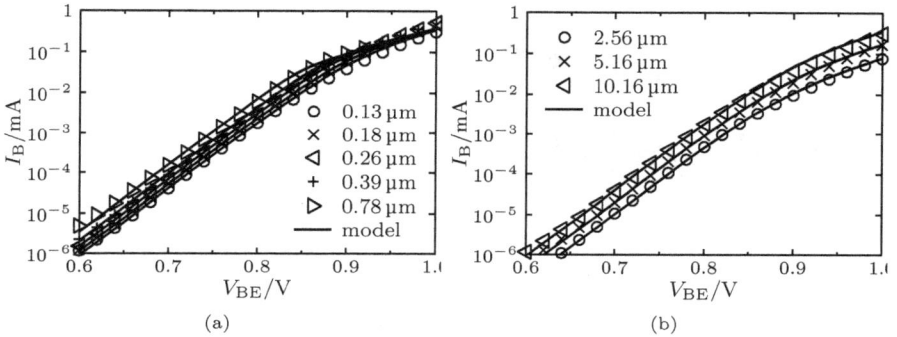

Figure B.9: Base current at $V_{BC} = 0\,V$ and room temperature for (a) a scaling versus emitter width and (b) emitter length.

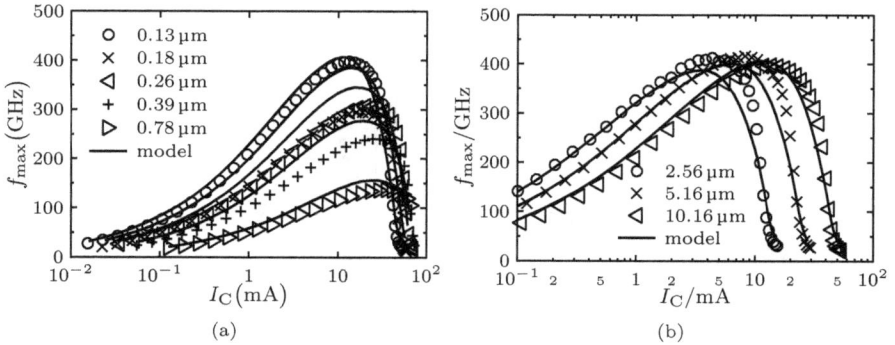

Figure B.10: Maximum oscillation frequency at $V_{BC} = 0\,V$ and room temperature for (a) a scaling versus emitter width and (b) emitter length and spot-frequency $f = 20\,GHz$.

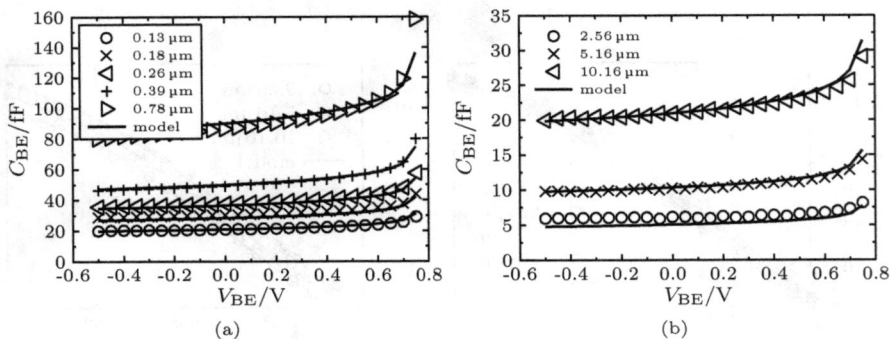

Figure B.11: Terminal BE-capacitance versus BE-voltage at $V_{CE} = 0\,V$ and room temperature for (a) a scaling versus emitter width and (b) emitter length and spot-frequency $f = 5\,GHz$.

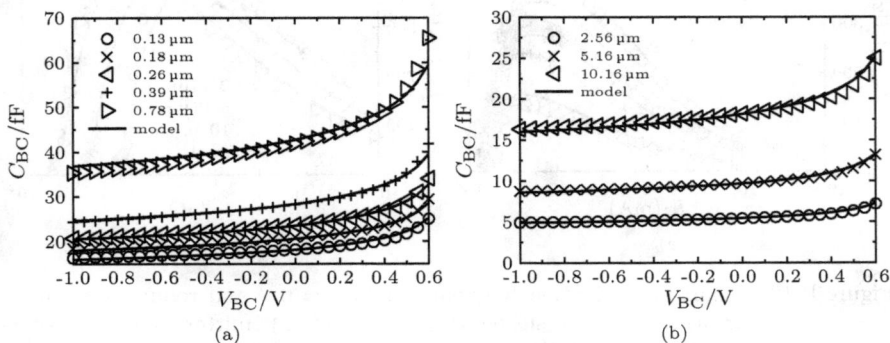

Figure B.12: Terminal BC-capacitance versus BC-voltage at $V_{BE} = 0\,V$ and room temperature for (a) a scaling versus emitter width and (b) emitter length and spot-frequency $f = 5\,GHz$.

B.3 Non-linear model verification

B.3.1 Number of harmonics in HB simulations

In HB simulations the number of harmonics are specified. The same holds for measurements with a non-linear network analyzer. For this equipment the number of
harmonics is limited by the available frequency range and the fundamental frequency.
One of the advantages of non-linear measurements and HB simulations is the possi-
bility to obtain time-domain curves, e.g., $i_C(t)$ or dynamic load lines, by inversion of
the Fourier series. In this section a short overview of the results from HB to transient
curves is given. For measurements often a number of five harmonics is included which
limits the fundamental frequency to roughly 10 GHz for existing equipment. During
the theoretical investigations simulations with the fundamental frequencies [10, 50,
200] GHz were performed. The goal here is to give an overview about the number
of harmonics that are required to result in reasonable agreement in time-domain.

Note that basically a simple Fourier series of the data from the time-domain
simulation would be sufficient for this comparison. However, as shown later even
with a sufficient number of harmonics HB simulations might give different results
compared to transient simulations. Moreover, the results in frequency-domain from
HB simulation depend on the number of harmonics as shown in Fig. B.13. There,
the output power at fundamental and second and third harmonic versus available
input power is given. Especially when only using three harmonics during simulation,
results strongly differ from results obtained when using a larger number.

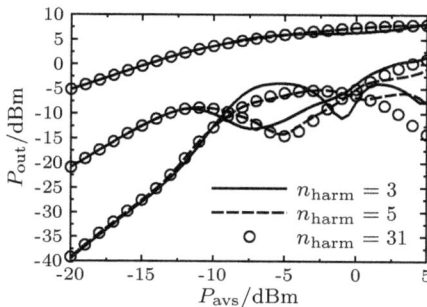

Figure B.13: Comparison of P_{out} for the first three harmonics obtained from HB
simulation with different number of harmonics (n_{harm}). Results are
given here for $f_0 = 10$ GHz.

Results from compact model simulation in time- and frequency-domain are given
in Fig. B.14-B.16. Shown for each frequency is an input power corresponding to
$P_{in,1dB}$ and an input power far in the compression region. The most notable result
is that for high input power HB does not provide the same results as transient simulation, although the shape is identical and the deviations are small. Obviously, for

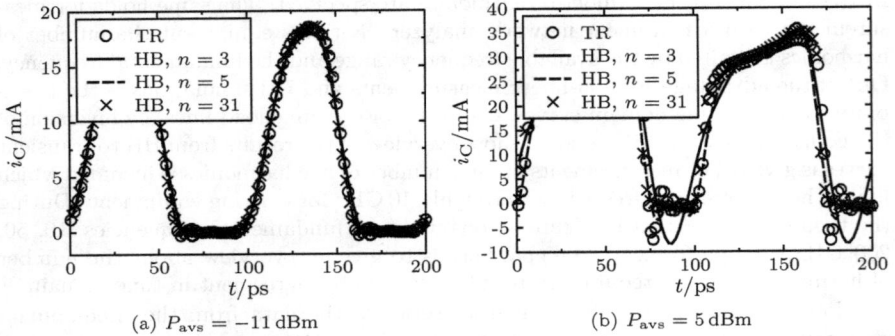

(a) $P_{\mathrm{avs}} = -11\,\mathrm{dBm}$ (b) $P_{\mathrm{avs}} = 5\,\mathrm{dBm}$

Figure B.14: Time domain $i_{\mathrm{C}}(t)$ at $f_0 = 10\,\mathrm{GHz}$ and different input power. For this frequency, $P_{\mathrm{avs}} = -11\,\mathrm{dBm}$ corresponds to $P_{\mathrm{in,1dB}}$.

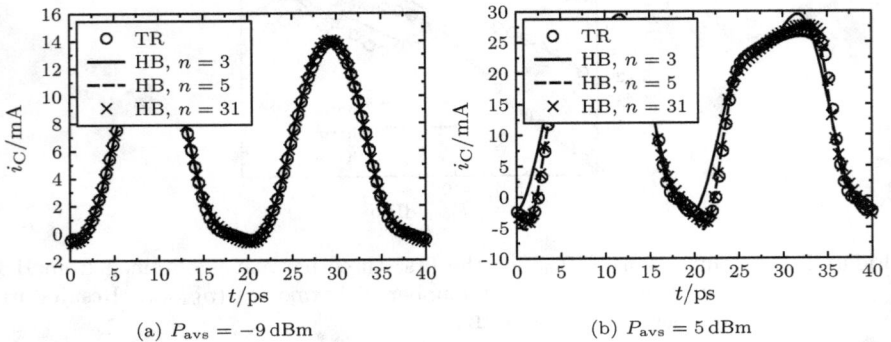

(a) $P_{\mathrm{avs}} = -9\,\mathrm{dBm}$ (b) $P_{\mathrm{avs}} = 5\,\mathrm{dBm}$

Figure B.15: Time domain $i_{\mathrm{C}}(t)$ at $f_0 = 50\,\mathrm{GHz}$ and different input power. For this frequency, $P_{\mathrm{avs}} = -9\,\mathrm{dBm}$ corresponds to $P_{\mathrm{in,1dB}}$.

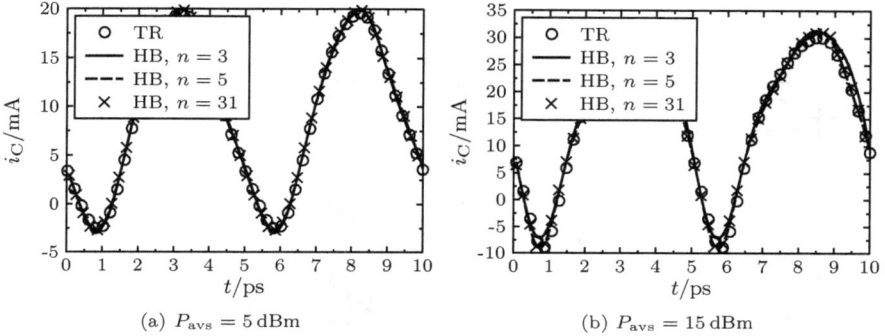

(a) $P_{\mathrm{avs}} = 5\,\mathrm{dBm}$ (b) $P_{\mathrm{avs}} = 15\,\mathrm{dBm}$

Figure B.16: Time domain $i_C(t)$ at $f_0 = 200\,\mathrm{GHz}$ and different input power. For this frequency, $P_{\mathrm{avs}} = 5\,\mathrm{dBm}$ corresponds to $P_{\mathrm{in,1dB}}$.

high input power the number of 3 or 5 harmonics is not sufficient to return accurate time-domain results. However, for high frequencies - 200 GHz in this case, i.e. close to f_T for the given operating point and model - the impact of the dynamic elements becomes significant and leads to a linearization due to the weak non-linearity of the capacitances and the larger dynamic voltage drop across the series resistances. Thus, for high frequencies a lower number of harmonics is sufficient to calculate reasonable time domain curves. However, for compact model simulation, using a larger number of harmonics should be preferred in order to obtain correct results.

B.3.2 Model verification

Figures B.17-B.20 provide a comparison of the measured I_C, P_{out}, G_T and PAE for $f_0 = 2.5\,\mathrm{GHz}$ and 10 GHz. The results for the latter frequency suffer from the effects discussed in 4.4.5.4.

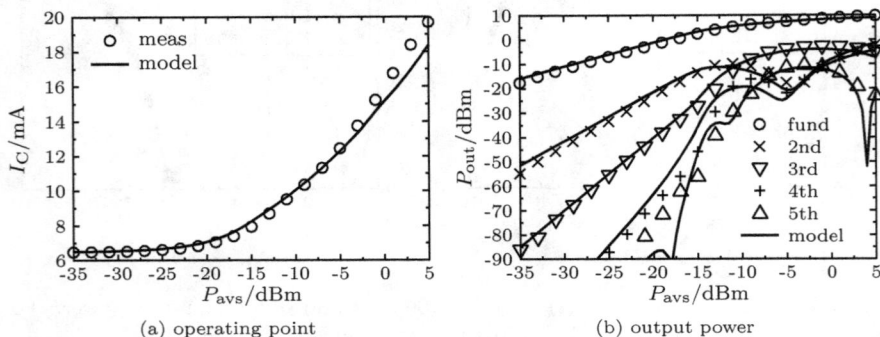

(a) operating point   (b) output power

Figure B.17: Operating point and fundamental and harmonic components of the output power as a function of available input power for the fundamental frequency $f_0 = 2.5\,\text{GHz}$.

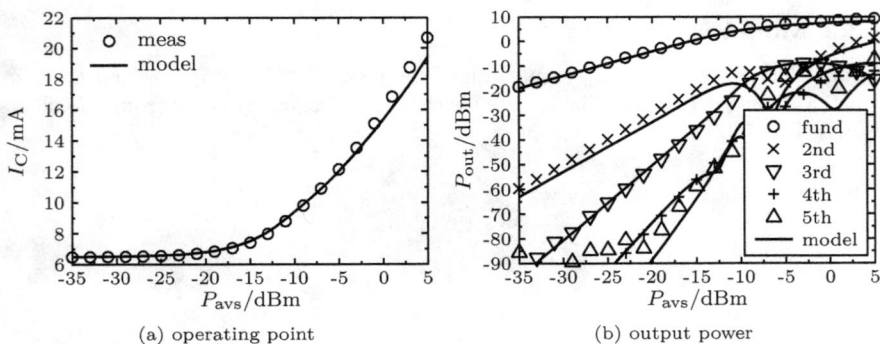

(a) operating point   (b) output power

Figure B.18: Operating point and fundamental and harmonic components of the output power as a function of available input power for the fundamental frequency $f_0 = 10\,\text{GHz}$.

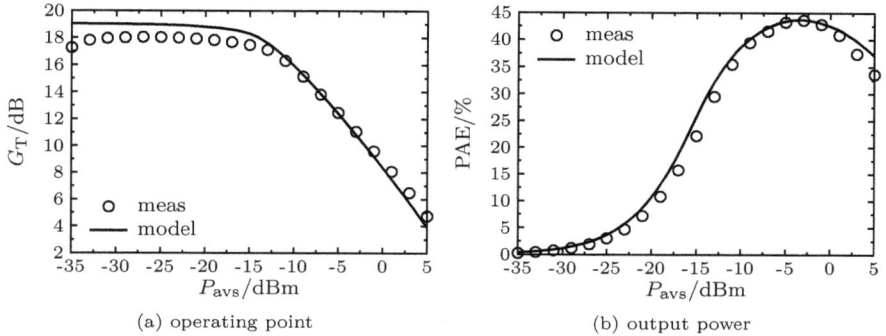

(a) operating point   (b) output power

Figure B.19: Transducer gain and PAE as a function of available input power for the fundamental frequency $f_0 = 2.5\,\text{GHz}$.

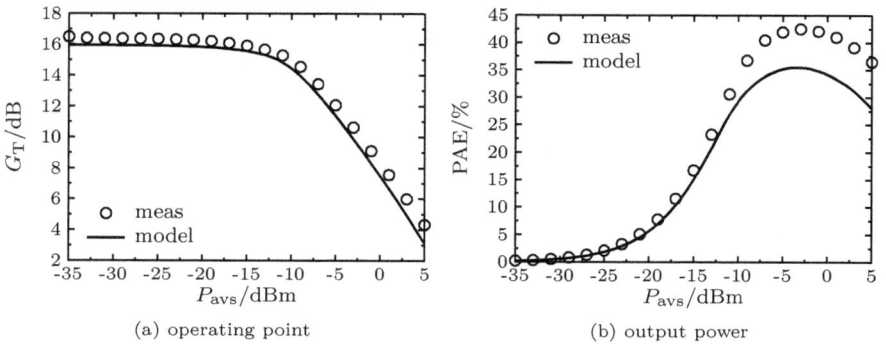

(a) operating point   (b) output power

Figure B.20: Transducer gain and PAE as a function of available input power for the fundamental frequency $f_0 = 10\,\text{GHz}$.

Resume

Andreas Pawlak

Date of birth: 04 November 1982
Place of birth: Köthen, Germany

Education and training

2007	Dresden Microelectronics Academy Dresden, Germany
10/2003 - 12/2008	Diplom-Ingenieur corresponding to Master of Science Technische Universität Dresden, Germany

- Electrical engineering, Microelectronics

1989 - 2002	A-level Gymnasium "An der Rüsternbreite", Köthen, Germany

- Exams: Mathematics, Physics, English, Geographics

Work experience

01/2009	Research assistant Technische Universität Dresden, Chair for Electron Devices and Integrated Circuits

- Modeling of advanced Silicon-Germanium Heterojunction Bipolar Transistors
- On-wafer DC and HF measurement and characterization of HBTs
- Development and support for the bipolar transistor compact models HICUM/L2 and HICUM/L0

10/2007 - 03/2008	Internship PHILIPS Medical Systems, Böblingen, Germany

- Development of a novel method for calibrating of touch screens for medical application

- Research on capacitive touch sensors

11/2005 - 09/2007   Student Research Assistant
Technische Universität Dresden, Chair for Electron Devices and Integrated Circuits

- Research on advances Silicon-Germanium based processes
- Simulation and modeling of electrical devices, focus on Heterojunction Bipolar Transistors

05/2005 - 09/2007   Student Assistant
Technische Universität Dresden, Centre for Information Technology Services and High Performance Computing

- Supervisor of PC-Pools
- Supply of data media

04/2003 - 06/2003   Pre-study industrial practical work
TACK GmbH, Köthen, Germany

- Construction and testing of distribution boxes and circuit boards

Dresden, October 25, 2015

List of publications

Journals

[1] S. Lehmann, Y. Zimmermann, A. Pawlak, and M. Schroter, "Characterization of the static thermal coupling between emitter fingers of bipolar transistors," *IEEE Transactions on Electron Devices*, vol. 61, no. 11, pp. 3676-3683, Nov 2014.

[2] A. Pawlak and M. Schroter "An Improved Transfer Current Model for RF and mm-Wave SiGe(C) Heterojunction Bipolar Transistors," *IEEE Transactions on Electron Devices*, vol. 61, no. 8, pp. 2612-2618, Aug 2014.

[3] F. Ellinger, T. Mikolajick, G. Fettweis, D. Hentschel, S. Kolodinski, H. Warnecke, T. Reppe, C. Tzschoppe, J. Dohl, C. Carta, D. Fritsche, G. Tretter, M. Wiatr, S. D. Kronholz, R. P. Mikalo, H. Heinrich, R. Paulo, R. Wolf, J. Hübner, J. Waltsgott, K. Meißner, R. Richter, O. Michler, M. Bausinger, H. Mehlich, M. Hahmann, H. Möller, M. Wiemer, H.-J. Holland, R. Gärtner, S. Schubert, A. Richter, A. Strobel, A. Fehske, S. Cech, U. Aßmann, A. Pawlak, M. Schröter, W. Finger, S. Schumann, S. Höppner, D. Walter, H. Eisenreich and R. Schüffny, "Energy efficiency enhancements for semiconductors, communications, sensors and software achieved in cool silicon cluster project," *The European Physical Journal - Applied Physics*, vol. 63, no. 01, pp. 14402-p1-p12, July 2013.

Conferences

[4] A. Pawlak, S. Lehmann, P. Sakalas, J. Krause, K. Aufinger, B. Ardouin, M. Schroter, "SiGe HBT modeling for mm-wave circuit design," in *IEEE Bipolar/BiCMOS Circuits and Technology Meeting (BCTM)*, pp. n/A, Oct. 2015.

[5] A. Pawlak, S. Lehmann, and M. Schroter, "A simple and accurate method for extracting the emitter and thermal resistance of BJTs and HBTs," in *IEEE Bipolar/BiCMOS Circuits and Technology Meeting (BCTM)*, pp. 175-178, Oct. 2014.

[6] A. Pawlak, M. Schroter, A. Fox, "Geometry Scalable Model Parameter Extraction for mm-Wave SiGe-Heterojunction Transistors," In *IEEE Bipolar/BiCMOS Circuits and Technology Meeting (BCTM)*, pp. 127-130, Oct. 2013.

[7] T. Rosenbaum, A. Pawlak, M. Schroter, S. Lehmann, "Automated transit time and transfer current extraction for single transistor geometries," In *IEEE Bipolar/BiCMOS Circuits and Technology Meeting (BCTM)*, pp. 25-28, Oct. 2013.

[8] F. Ellinger, G. Fettweis, C. Tzschoppe, C. Carta, D. Fritsche, G. Tretter, U. Yodprasit, R. Paulo, A. Richter, A. Strobel, R. Wolf, A. Fehske, C. Isheden, A. Pawlak, M. Schroter, S. Schumann, S. Hoppner, D. Walter, H. Eisenreich, R. Schuffny, "Power-efficient high-frequency integrated circuits and communi-

cation systems developed within Cool Silicon cluster project," In *IEEE MTT-S International Microwave & Optoelectronics Conference (IMOC)*, pp. 1-2, Aug. 2013.

[9] M. Schroter, S. Chaudhry, J. Zheng, A. Mukherjee, A. Pawlak, S. Lehmann, "SiGe HBT compact modeling for production-type circuit design," In *IEEE 12th Topical Meeting on Silicon Monolithic Integrated Circuits in RF Systems (SiRF)*, pp. 129-132, Jan. 2012.

[10] S. Lehmann, M. Weiss, Y. Zimmermann, A. Pawlak, K. Aufinger, M. Schroter, "Scalable compact modeling for SiGe HBTs suitable for microwave radar applications," In *IEEE Topical Meeting on Silicon Monolithic Integrated Circuits in RF Systems (SiRF)*, pp. 113-116, Jan. 2011.

[11] M. Schroter, A. Pawlak, P. Sakalas, J. Krause, T. Nardmann, "SiGeC and InP HBT Compact Modeling for mm-Wave and THz Applications," In *IEEE Compound Semiconductor Integrated Circuit Symposium (CSICS)*, pp. 1-4, Oct. 2011.

[12] A. Pawlak, M. Schroter, J. Krause, D. Céli, N. Derrier, "HICUM/L2 v2.3 parameter extraction for advanced SiGe-heterojunction bipolar transistors," In *IEEE Bipolar/BiCMOS Circuits and Technology Meeting (BCTM)*, pp. 195-198, Oct. 2011.

[13] A. Pawlak, M. Schroter, A. Mukherjee, S. Lehmann, S. Shou, D. Céli, "Automated model complexity reduction using the HICUM hierarchy," In *Semiconductor Conference Dresden (SCD)*, pp 1-4, Sept. 2011.

[14] B. Ardouin, C. Raya, M. Schroter, A. Pawlak, D. Céli, F. Pourchon, K. Aufinger, T.F. Meister, T. Zimmer, "Modeling and parameter extraction of SiGe: C HBT's with HICUM for the emerging terahertz era," In *European Microwave Integrated Circuits Conference (EuMIC)*, pp. 25-28, Sept. 2010.

[15] A. Pawlak, M. Schroter, J. Krause, G. Wedel, C. Jungemann, "On the Feasibility of 500 GHz Silicon-Germanium HBTs," In *International Conference on Simulation of Semiconductor Processes and Devices (SISPAD)*, pp .1-4, Sept. 2009.

Workshops

[16] M. Schroter, A. Pawlak, P. Sakalas, "Large-signal nonlinear HBT compact modeling and verification for mm-wave applications", *IEEE Top. Symposium on Power Amplifiers for Wireless communications*, San Diego, USA, 2015.

[17] A. Pawlak, ""Modeling of physical effects in bipolar transistors for mm-wave circuits", *ESSCIRC, Tutorial: RF Design in BiCMOS*, Graz, Austria, 2015.

[18] P. Sakalas, A. Pawlak, M. Schroter, "Nonlinear distortion in mm-wave SiGe HBTs: modeling and measurements", *DOT7, THz -Workshop: Millimeter and Sub-Millimeter wave circuit design and characterization*, Venice, 2014.

[19] M. Schröter, A. Pawlak, "Physics-based nonlinear compact modeling of HBTs for mm-wave applications", *International Microwave Symposium*, Tampa Bay, Florida, USA, 2014.

[20] P. Sakalas, A. Pawlak, S. Lehmann, "Nonlinear characterization and modeling of advanced high speed SiGe HBTs and InP DHBTs", *International Microwave Symposium*, Tampa Bay, Florida, USA, 2014.

[21] A. Pawlak, M. Schröter, A. Mukherjee, "Status of HICUM/L2 Model", *Bipolar Arbeitskreis*, Crolles, France, 2014.

[22] A. Pawlak, M. Schröter, A. Mukherjee, "Status of HICUM/L2 Model", *14th HICUM Workshop*, RFMD, Greensboro (NC), USA, 2014.

[23] A. Pawlak, M. Schroter, S. Lehmann, "Coupled extraction of RE and RTH based on DC output curves", *Bipolar Arbeitskreis*, Frankfurt (Oder), Germany, 2013.

[24] A. Pawlak, M. Schroter, A. Fox, "Geometry Scalable Model Parameter Extraction for mm-Wave SiGe-Heterojunction Transistors", *13th HICUM Workshop*, Delft, Netherlands, 2013.

[25] A. Mukherjee, A. Pawlak, M. Schröter, "Status of HICUM/L0 Model", *13th HICUM Workshop*, Delft, Netherlands, 2013.

[26] A. Pawlak, M. Schroter, A. Mukherjee, J. Krause, "Modeling high-speed SiGe-HBTs with HICUM/L2 v2.31", *12th HICUM Workshop*, Newport Beach, CA (USA), 2012.

[27] A. Pawlak, S. Lehmann, P. Ehrlich, M. Schroter, "Modeling the collector related charge in SiGe-HBTs", *12th HICUM Workshop*, Newport Beach, CA (USA), 2012.

[28] A. Pawlak, M. Schröter, A. Mukherjee, S. Lehmann, "Automated parameter conversion from HICUM/L2 to HICUM/L0", *11th HICUM Workshop*, Bordeaux, France, 2011.

[29] A. Pawlak, M. Schröter, "Application of HICUM/L2 v2.30 to advanced multi-100GHz SiGe HBTs", *Bipolar Arbeitskreis*, Crolles, France, 2010..

[30] A. Pawlak, M. Schröter, A. Mukherjee, T. Kessler, "HICUM/L2 v2.30 overview", *10th HICUM Workshop*, Dresden, Germany, 2010.

[31] A. Pawlak, M. Schröter, J. Krause, "A HICUM extension for medium current densities", *9th HICUM Workshop*, Würzburg, Germany, 2009.